AF822065

My Book

This book belongs to

Cover Design by :
Gowri Vemuri

First Edition :
May, 2019

Author :
Gowri Vemuri

Edited by :
Raksha Pothapragada
Ritvik Pothapragada

Questions: mathknots.help@gmail.com

NOTE : VDOE is neither affiliated nor sponsors or endorses this product.

Dedication

This book is dedicated to:

My Mom, who is my best critic, guide and supporter.

To what I am today, and what I am going to become tomorrow,

is all because of your blessings, unconditional affection and support.

This book is dedicated to the

strongest women of my life ,

my dearest mom

and

to all those moms in this universe.

G.V.

The Virginia Board of Education and Virginia Department of Education (VDOE) have developed the Virginia Assessment Program (VAP) to measure and evaluate students' academic progress in the Standards of Learning (SOLs). The SOLs indicate Virginia's expectations for what students should know and be able to do in the subject areas of reading, writing, mathematics, science, and history/social science.

Students in grades 3-12 to take the Standards of Learning (SOL) assessments each year. Some of the tests are required by all students each year, and others are required only at specific grade levels. Additionally, with the removal of some SOL tests in recent years, the VDOE assigned the responsibility of the creation and administration of alternate, performance - based assessments on local divisions. Student scores from these tests determine a school's and the division's state accreditation and measures progress toward meeting federal targets.

Virginia Standards of Learning (SOL) tests are generally given online unless a student has an identified and documented need to be assessed using paper,pencil format. The test question format is typically multiple choice, and each test contains some technology enhanced items.

GRADE 3-8 STANDARDS OF LEARNING(SOL) TESTS

GRADE 3	GRADE 4	GRADE 5	GRADE 6	GRADE 7	GRADE 8
	VIRGINIA STUDIES				WRITING
MATH	MATH	MATH	MATH	MATH	MATH
READING	READING	READING	READING	READING	READING
		SCIENCE			SCIENCE

NOTE : VDOE is neither affiliated nor sponsors or endorses this product.

END OF COURSE STANDARDS OF LEARNING (SOL) TESTS

GRADE 9	GRADE 10	GRADE 11
ALGEBRA I	GEOMETRY	ALGEBRA II
EARTH SCIENCE	BIOLOGY	CHEMISTRY
WORLD HISTORY I	WORLD HISTORY II	VIRGINIA & US HSTORY
		WORLD GEOGRAPHY
		ENGLISH : READING
		ENGLISH : WRITING

Any Student taking one of the courses listed here is expected to take the corresponding end-of-course SOL test. The grade levels depicted here represent grade level at which students typically participate in these courses.

NOTE : VDOE is neither affiliated nor sponsors or endorses this product.

SOL Test Scoring and Performance Reports:

Standards of Learning assessments in English reading, mathematics, science and history/social science are made up of 35-50 items or questions that measure content knowledge, scientific and mathematical processes, reasoning and critical thinking skills. English writing skills are measured with a two-part assessment that includes multiple-choice items and an essay.

Student performance is graded on a scale of 0-600 with 400 representing the minimum level of acceptable proficiency and 500 representing advanced proficiency. On English reading and mathematics tests, the Board of Education has defined three levels of student achievement: basic, proficient, and advanced, with basic describing progress towards proficiency.

Performance Achievement Levels:

- The achievement levels for grades 3-8 reading and mathematics tests are: *Pass/Advanced*, *Pass/Proficient*, *Fail/Basic*, and *Fail/Below Basic*.

- The achievement levels for science tests, history tests, and End-of-Course (EOC) tests* are: *Pass/Advanced*, *Pass/Proficient*, and *Fail/Does Not Meet*.

- The EOC Writing (2010 SOL) test, EOC Reading (2010 SOL) test, and EOC Algebra II (2009 SOL) test have an achievement level of *Advanced/College Path* in place of the *Pass/Advanced* achievement level.

NOTE : VDOE is neither affiliated nor sponsors or endorses this product.

INDEX

FORMULA SHEET

1. Area of a triangle

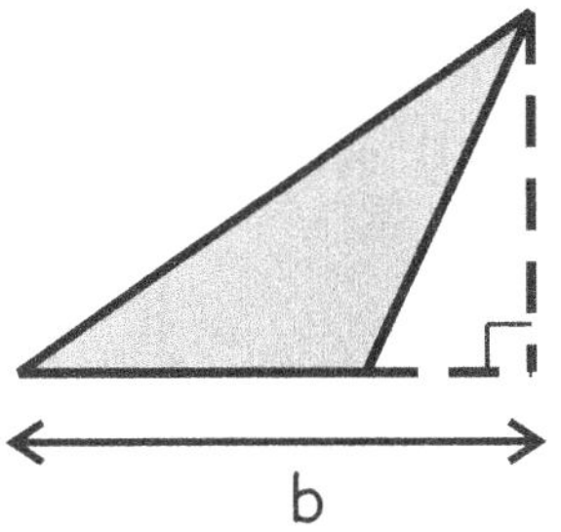

$A = \frac{1}{2} bh$

2. Area of a parellelogram

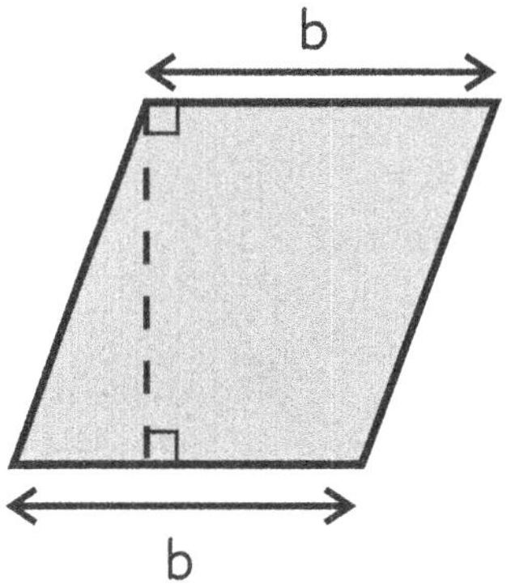

$A = bh$

3. Volume and Surface area of a cuboid

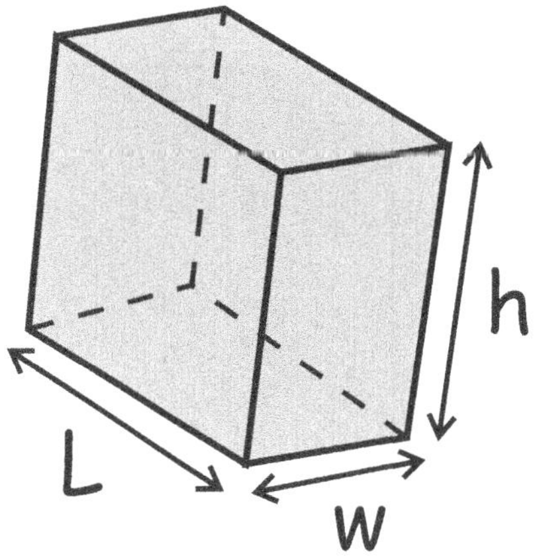

$V = lwh$

$S.A = 2(lw + lh + wh)$

4. Perimeter and Area of a Square

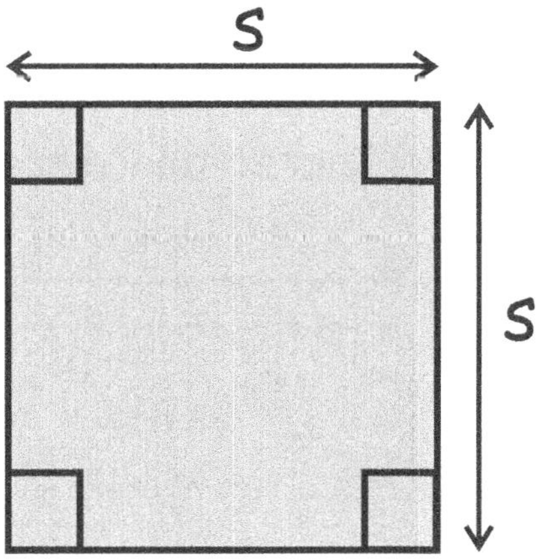

$P = 4s$

$A = s^2$

5. Area of a Trapezium

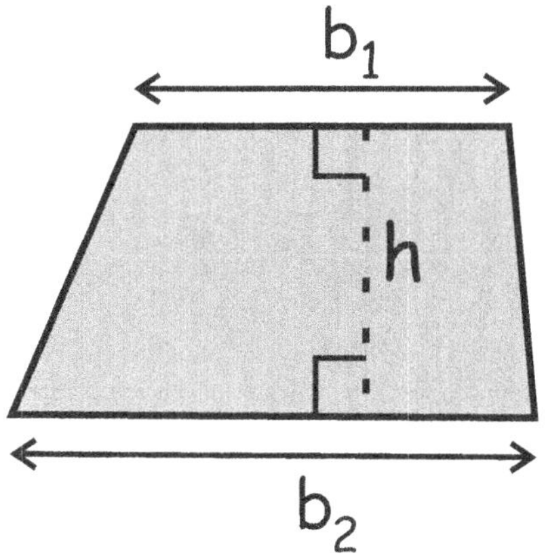

$A = \frac{1}{2} h(b_1 + b_2)$

6. Circumference and Area of a Circle

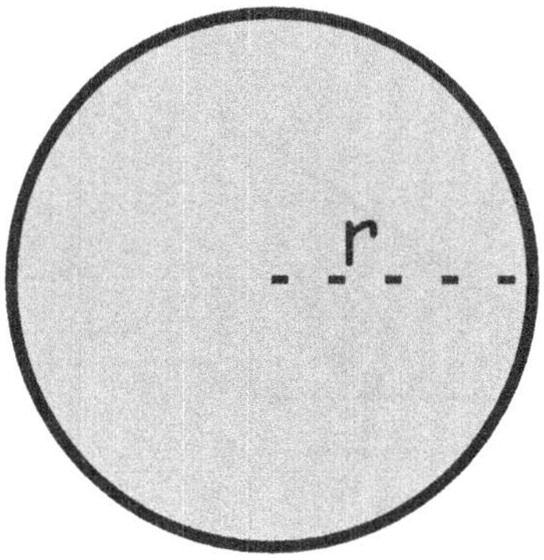

$c = 2\pi r$

$A = \pi r^2$

pi

$\pi = 3.14$

$\pi = \frac{22}{7}$

7. Perimeter and Area of a Rectangle

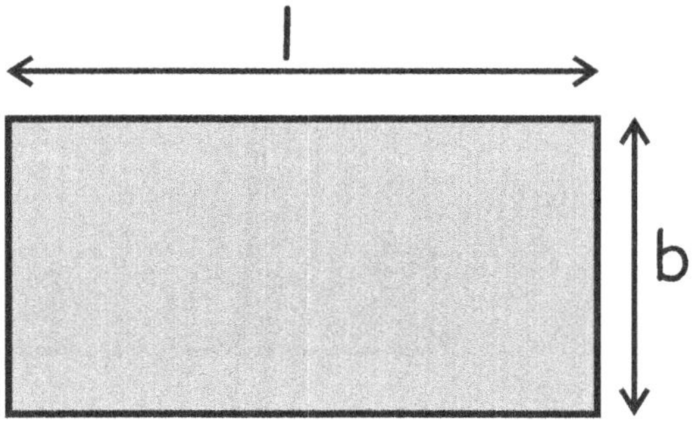

Area = l b ×

Perimeter = 2(l + b)

8. Acute angle

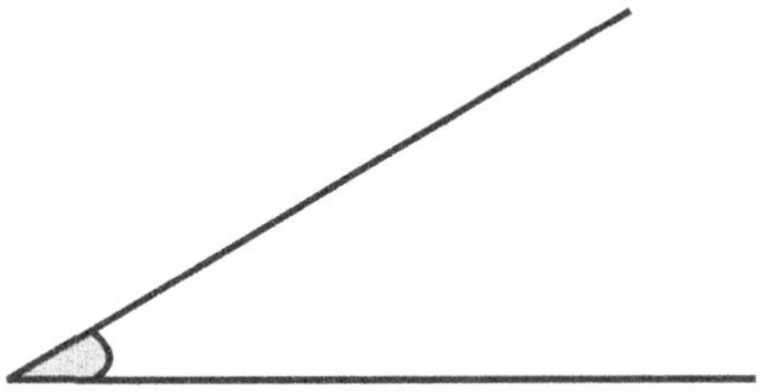

Any angle less than 90^0
is called as acute angled

9. Right angle

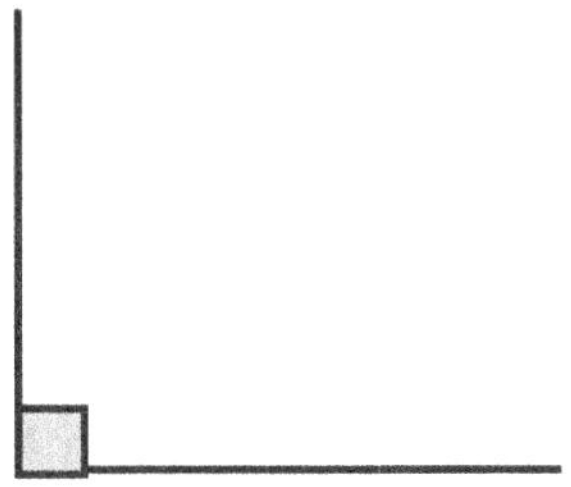

An angle equal to 90^0
is called as right angled

10. Obtuse angle

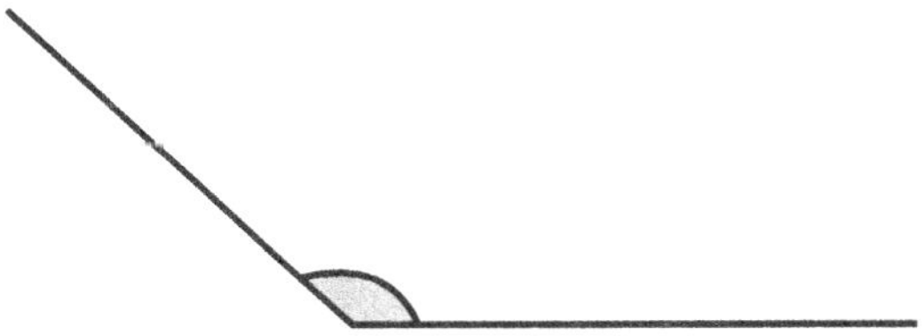

Any angle greater than 90^0
is called as obtuse angled

11. Perimeter of a polygon KLMNOP

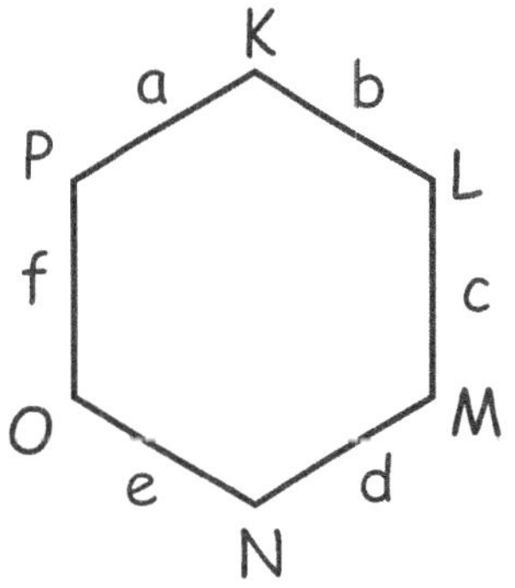

Perimeter =
(a + b + c + d + e + f) units

Abbreviations

milligram	mg
gram	g
kilogram	kg
milliliter	mL
liter	L
kiloliter	kL
millimeter	mm
centimeter	cm
meter	m
kilometer	km
square centimeter	cm^2
cubic centimeter	cm^3

year	yr
month	mon
hour	hr
minute	min
second	sec

volume	V
total Square Area	S.A
area of base	B
ounce	oz
pound	lb
quart	qt
gallon	gal.
inches	in.
foot	ft
yard	yd
mile	mi.
square inch	sq in.
square foot	sq ft
cubic inch	cu in.
cubic foot	cu ft

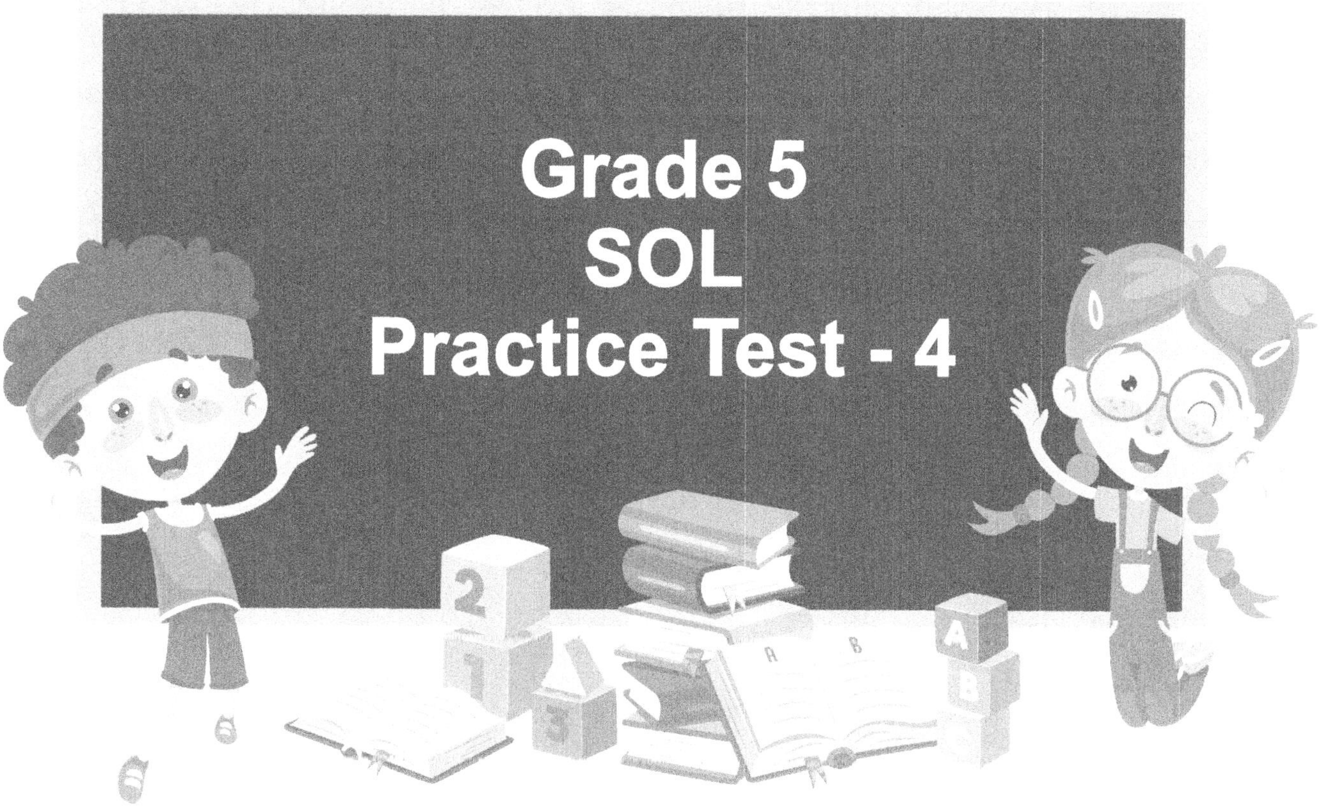
Grade 5
SOL
Practice Test - 4

1. Evaluate

$$19.30 \div 2 = \boxed{?}$$

(A) 9.65 (B) 8.60

(C) 9.55 (D) 8.65

2. Maria saves at least $18 every week from her pay check. What is the least number of money she could have saved after 25 weeks ?

(A) 444 (B) 398 (C) 450 (D) 464

3. What is the sum of $5\frac{1}{4} + 1\frac{3}{4}$?

(A) 6 (B) $6\frac{3}{4}$

(C) $4\frac{3}{4}$ (D) 7

4. Tiny Tots school music department raised $4125 funds last year. This year they have raised $3197. How much more did they raise last year than this year ?

(A) 999 (B) 929

(C) 928 (D) 989

5. Evaluate $9001 \div 9 = \boxed{?}$

(A) 1000 R 1

(B) 1000 R 9

(C) 9000 R 9

(D) 9000 R 1

6. Evaluate

```
  421.324
+ 138.513
---------
```

(A) 282.811

(B) 559.837

(C) 283.881

(D) 559.811

7. Cathy slices the cake into 16 pieces and distributes it among 8 kids. How many pieces of cake does each child will eat ?

(A) 2 (B) 5 (C) 8 (D) 4

8. $\frac{1}{3} - \frac{1}{5} = ?$

(A) $\frac{4}{7}$

(B) $\frac{5}{15}$

(C) $\frac{2}{15}$

(D) $\frac{3}{15}$

9. Evaluate $6137 \div 6 = \boxed{?}$

(A) 1022 R 0 (B) 1022 R 5 (C) 1021 R 3 (D) 1022 R 2

10. Estimate the value of 9.0511 + 1. 0011

(A) 10.05
(B) 11.1522
(C) 8.539
(D) 10.0522

11. The Super fast shipping company delivers 3133 , 2121 ,1090 and 4100 packages on December 21st , 22nd ,23rd and 24th of the year 2018 .
How many packages are delivered in total from December 21st through 24th.

(A) 10444
(B) 10446
(C) 10448
(C) 10244

12. Order the below fractions from least to greatest ?

$$\frac{5}{8}, \frac{5}{4}, \frac{1}{2}, \frac{9}{16}$$

(A) $\frac{5}{8}, \frac{5}{4}, \frac{1}{2}, \frac{9}{16}$
(B) $\frac{9}{16}, \frac{1}{2}, \frac{5}{4}, \frac{5}{8}$
(C) $\frac{1}{2}, \frac{9}{16}, \frac{5}{8}, \frac{5}{4}$
(D) $\frac{5}{8}, \frac{9}{16}, \frac{5}{4}, \frac{1}{2}$

13. What is the place value of 7 in 401.3217 ?

(A) One - thousandths (B) Tenths

(C) Hundredths (D) Ten - thousandths

14. Write 0.761 in words ?

(A) Seven hundred sixty-one thousandths
(B) Seven and sixty-one thousandths
(C) Seven hundred sixty-one hundredths
(D) Seven hundred sixty-one

15. $\frac{74}{100}$ can also be written as ?

(A) 0.74 (B) 0.0074

(C) 7.04 (D) 0.074

16. Which number is five thousand one and one thousandths ?

(A) 5100.1 (B) 5001.000

(C) 5001.001 (D) 50001.01

17. Which of the below is correct ?

(A) $0.101 > 0.110$
(B) $0.111 > 1.111$
(C) $0.011 < 0.0011$
(D) $0.011 < 0.101$

18. Find the value of 3 in 76.32 ?

(A) 3 tens (B) 3 ones

(C) 3 tenths (D) 3 hundredths

19. Round 77.33 to nearest tenth ?

(A) 77.00 (B) 77.3

(C) 78.3 (D) 77.4

20. Which of the below angle measures 45^0 ?

(A) 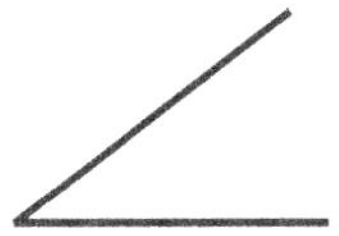

(B) 

(C) 

(D) 

21. Jackie has a vegetable garden in the form of pentagon with equal sides of length 13 meters. what is the perimeter of the vegetable garden ?

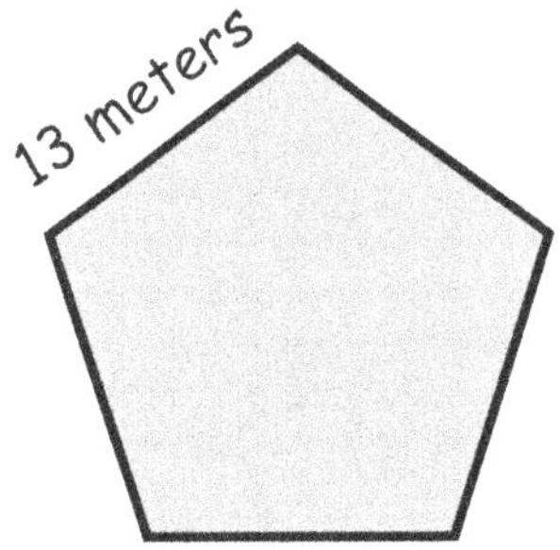

(A) 60 meters

(B) 65 meters

(C) 169 sq.meters

(D) 78 meters

22. What types of angles are used to form the below triangle ?

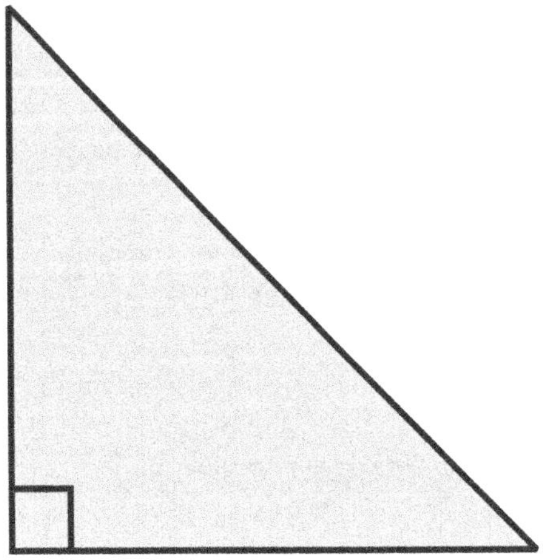

(A) Acute angle , Obtuse angle , Acute angle
(B) Acute angle , Right angle , Acute angle
(C) Acute angle , Acute angle , Acute angle
(D) Acute angle , Straight angle , Obtuse angle

23. Jack and Jill went up the hill trekking. They left at 7:23 a.m. They reached top of the hill by 6.55 pm ,by taking a break for 30 minutes for lunch. They camped on the top of the hill for the night.
What is the actual trekking time it took to reach the top of the hill ?

(A) 11 hours , 02 minutes
(B) 11 hours , 20 minutes
(C) 9 hours , 29 minutes
(D) 11 hours , 58 minutes

24. Find the area of the square plot below with the side length of 25 miles ?

(A) 100 miles
(B) 49 sq.miles
(C) 625 Sq.miles
(D) 50 miles

25. Which of the below can be measured in pounds ?

(A) Strawberries
(B) Milk
(C) Bus
(D) Juice Can

26. Which of the below picture shows the diameter AB ?

(A)

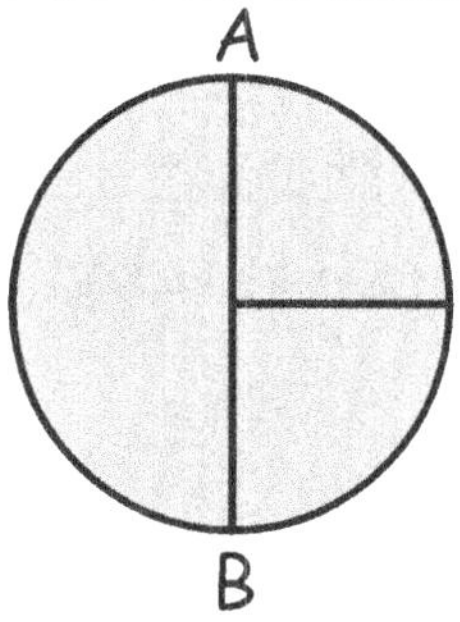

(B)

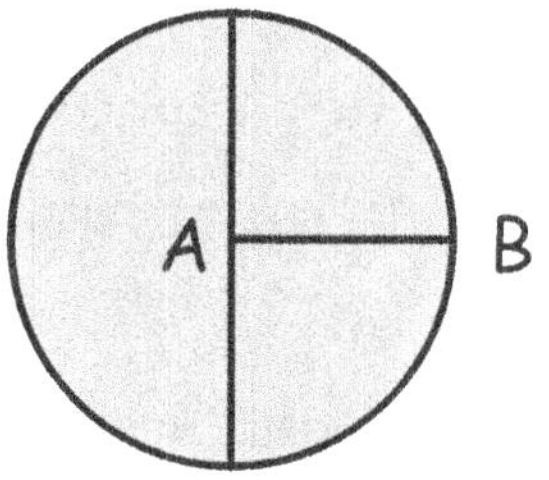

(C)

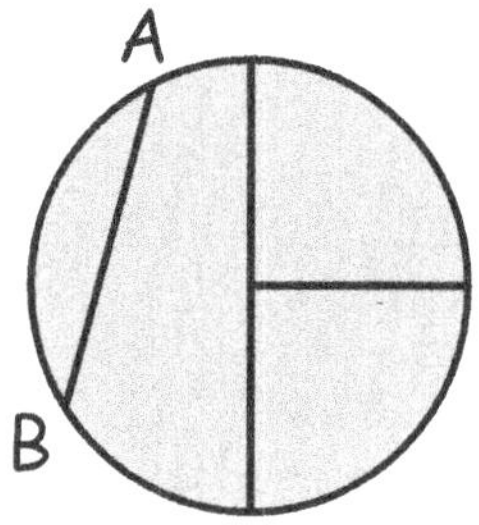

(D)

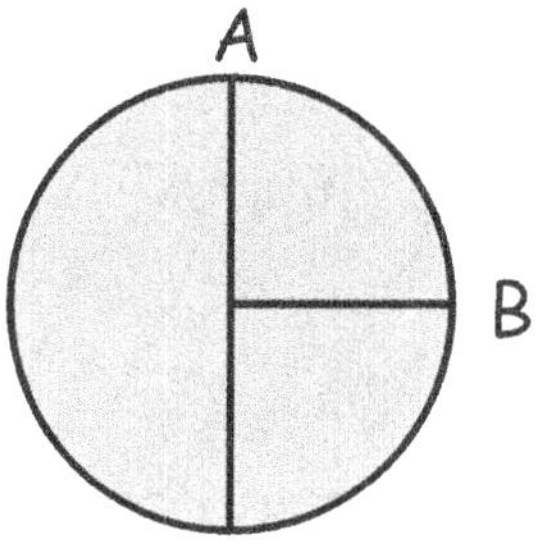

27. Which of the below picture has zero vertices and two edges ?

(A)

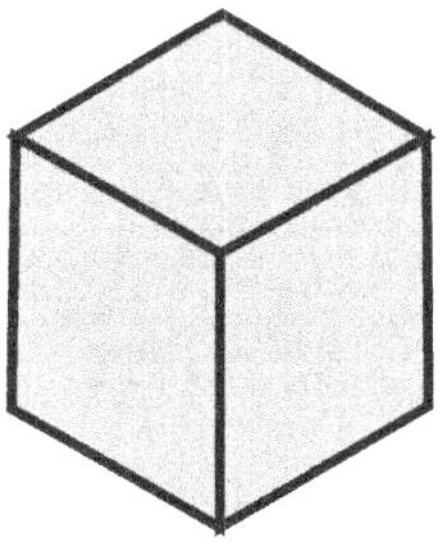

(B)

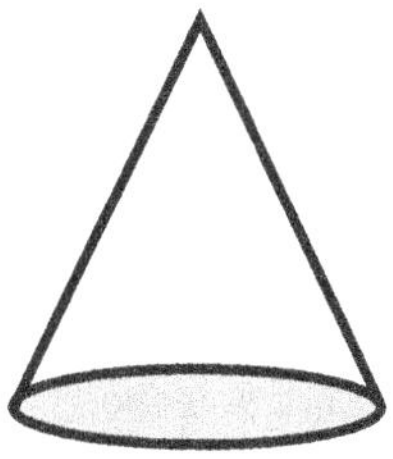

(C)

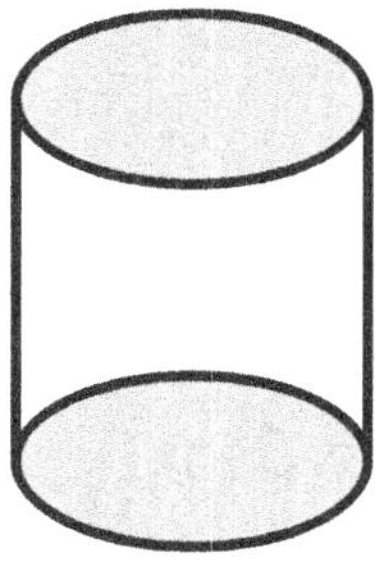

(D)

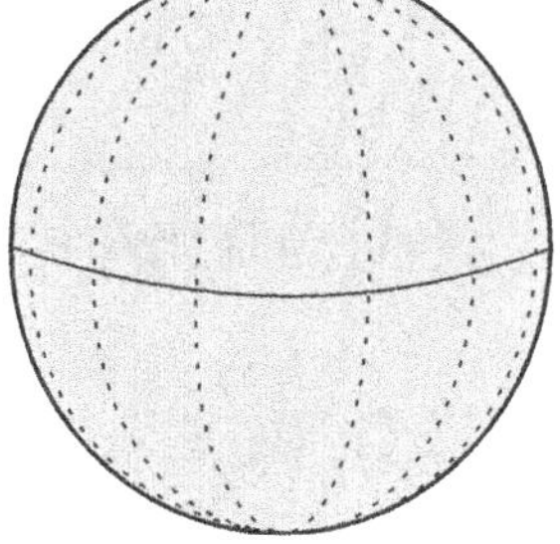

28. Which of the below pair of pictures are symmetric to each other ?

(A)

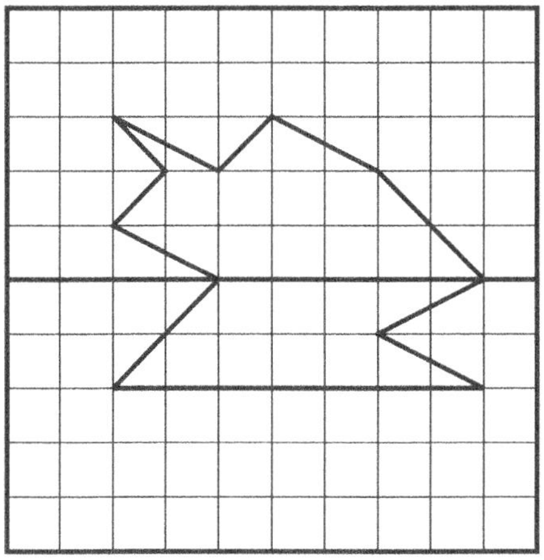

(B)

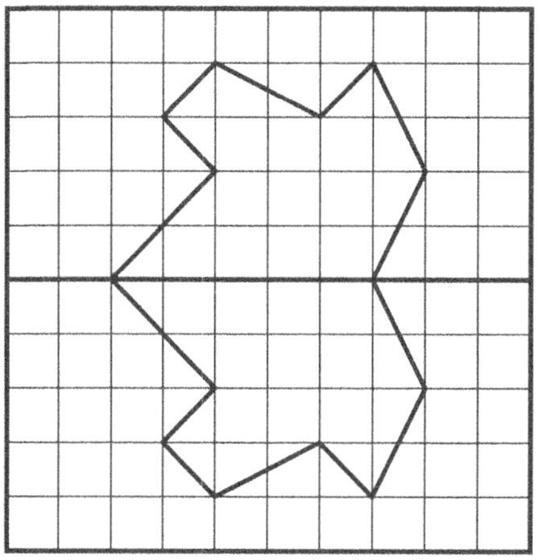

(C)

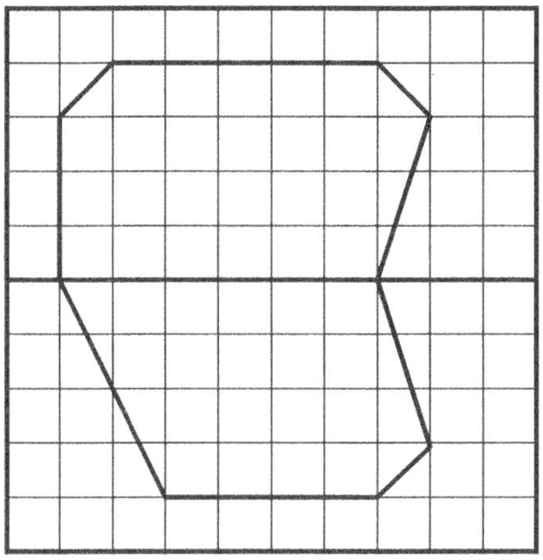

(D)

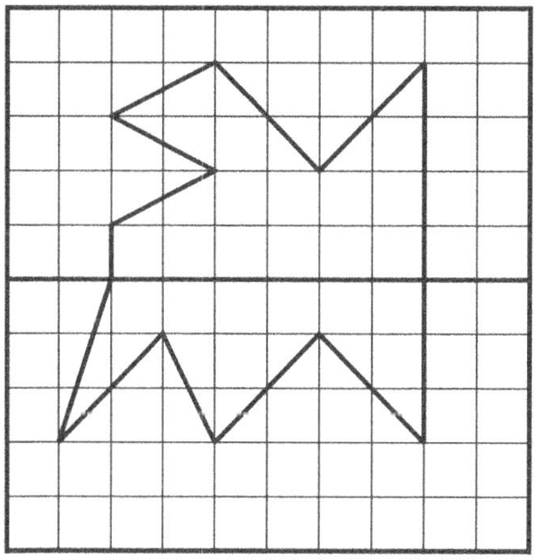

29. What is the probability of getting an odd number when a 8 sided dice is rolled ?

(A) $\frac{1}{2}$

(B) $\frac{1}{8}$

(C) $\frac{3}{8}$

(D) $\frac{5}{7}$

30. The Math quiz scores of fifth grade class are shown as stem-and-leaf plot as below. Based on the scores below, how many students have scored above 90 points

Stem	Leaf
6	1 , 4 , 6 , 7 , 8 , 8
7	4 , 6 , 7
8	1 , 2 , 3 , 9 , 5
9	0 , 0 , 3 , 5 , 7 , 8 , 9

Key	76
7	6

(A) 7   (B) 0   (C) 9   (D) 5

31. Below is sales data of various sandwiches at Deli stores this week. What is the mode of the data ?

19 , 11 , 10 , 9 , 9 , 0 , 7 , 3 , 9

(A) 9   (B) 6

(C) 11   (D) 3

32. The RR high school students voted for their one favorite sport. The results are shown below

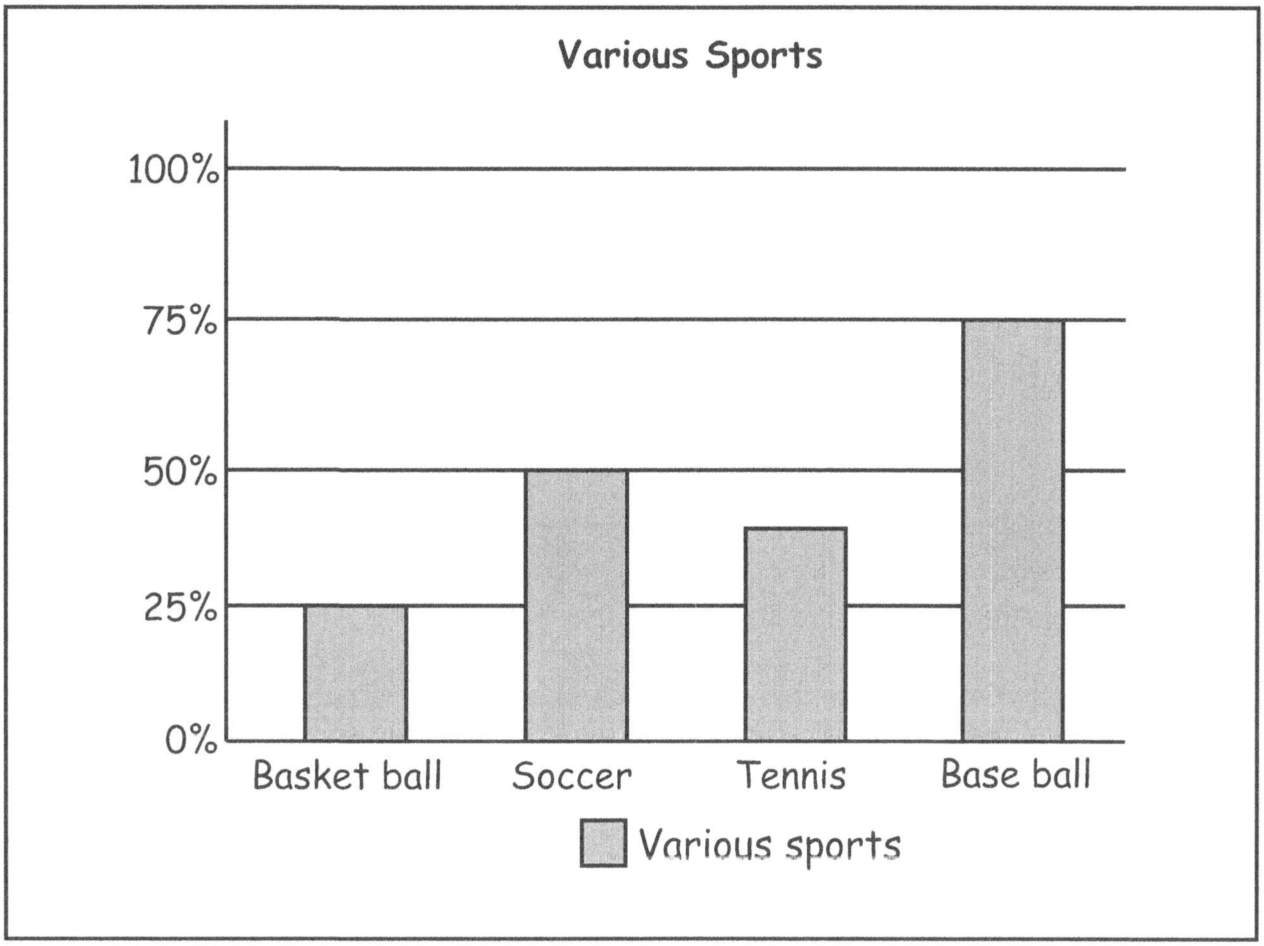

The greatest difference in the number of votes received was between the

(A) Soccer and Basket Ball

(B) Base Ball and Tennis

(C) Soccer and Tennis

(D) Base Ball and Basket Ball

33. Mrs.Sally is a pre-K teacher at Tiny Tots school. She places various shapes in a surprise box. Each student picks 2 shapes without looking at them. Based on the data in the table which of the below combinations is a possibility.

Counting Chips

Color	Number in Box
Circle	1
Triangle	9
Oval	1
Square	15
Hexagon	1

(A) 2 Triangles (B) 1 Oval, 2 Hexagons

(C) 2 Ovals (D) 2 Hexagons

34. Mary spins the arrow on the spinner shown.

What is the probability that the arrow will point to the section containing moons on the first try ?

(A) $\frac{1}{8}$ (B) $\frac{1}{4}$

(C) $\frac{3}{7}$ (D) $\frac{3}{4}$

35. what is the variable in the below expression ?

$$2d + 34 = 29$$

(A) 2 (B) 34 (C) d (D) X

36. Find the missing number in the pattern ?

7 , 12 , 17 , 22 , 27 , ? , 37 , 42

(A) 5 (B) 32 (C) 47 (D) 34

37. A number machine uses a rule to change numbers. The table shows the results.

In (p)	Out (q)
20	4
30	6
45	9
60	12

(A) $\frac{p}{5} = q$ (B) $p - 16 = q$

(C) $P.4 = q$ (D) $\frac{p}{4} = q$

38. Beth Knits three more scarfs than her friend Preet. If 's' represents the scarfs made by Preet, which of the below equation represents, the number of scarfs made by Beth.

(A) $s + 3$ (B) $S - 3$ (C) $\frac{S}{3}$ (D) $s \times 3$

39. Emma has a magic machine. When Emma inputs the number in the Triangle, a different number comes out to square based on a rule. Find the math rule that magic machine uses.

Input	Output
2	6
3	9
4	12

(A) Add 2 (B) Multiply by 2

(C) Subtract 1 (D) Multiply by 3

40. Find the missing number in the series

96 , 48 , 24 , ? , 6 , 3

(A) 10 (B) 22

(C) 12 (D) 16

41. Jacob works at a gas station for \$11 per hour during weekends. If he earns \$121 this weekend. What does "h" represent in the equation

$11h = 121$

(A) The number of hours Jacob worked last week

(B) The number of days Jacob worked last week

(C) The number of weeks Jacob worked last month

(D) The number of days Jacob worked last month

42. The three side lengths of an isosceles triangle are m, n and m. What equation can be used to find p, the perimeter of the triangle ?

(A) $p = 2m + 2n$ (B) $p = m + 2n$

(C) $p = m + n$ (D) $p = 2m + n$

43. Write the problem for the below expression ?

$$5X = 20$$

(A) Jason gives 20 candies to X students, a total of 5 candidates all together

(B) Jason has 20 candies. She puts them in 5 bags. If each bag has the same number of candies. How many candies are there in each bag ?

(C) Jason gives 10 candies to X students.

(D) Jason has 20 candies. He puts them in X bags. If each bag has X + 5 candies. How many are left ?

44. Which of the following can be solved using the open sentence below ?

$$p - 5 = s \text{ ; where } s = 7$$

(A) Pam has 5 more scarfs than Stella. If 's' represents the number of scarfs Stella has, how many scarfs does Pam have ?

(B) Pam has 5 times scarfs than Stella. If 's' represents the number of scarfs Stella has, how many scarfs does Pam have ?

(C) Pam has 5 fewer scarfs than Stella. If 's' represents the number of scarfs Stella has, how many scarfs does Pam have ?

(D) Stella has 5 times scarfs than Pam. If 's' represents the number of scarfs Stella has, how many scarfs does Stella have ?

45. Three high school juniors, Rita, Matt, and Tom are running for student council president. A survey is taken a week before the election asking 40 students which candidate, they will vote for in the election. The results are shown in the table below.

Contestant name	No. of students supporting
Rita	15
Mat	13
Tom	12

Based on the table, what is the probability that a student will vote for Rita?

(A) $\frac{3}{8}$

(B) $\frac{13}{40}$

(C) $\frac{12}{40}$

(D) $\frac{5}{8}$

46. Which set of data can be classified as quantitative ?
(A) Favorite desserts that you like
(B) Delegates at U.N.O
(C) Sales tax rate in each Virginia county
(D) Opinion of a middle School students on a favorite sport

47. Lilly made two bead necklaces of length 16.2 inches and 22.8 inches. What is the difference between the longest to shortest necklaces she made ?

(A) 66.6

(B) 6.3

(C) 6.5

(D) 6.6

48. Find the reflection of the first figure across the dashed line ?

(A)

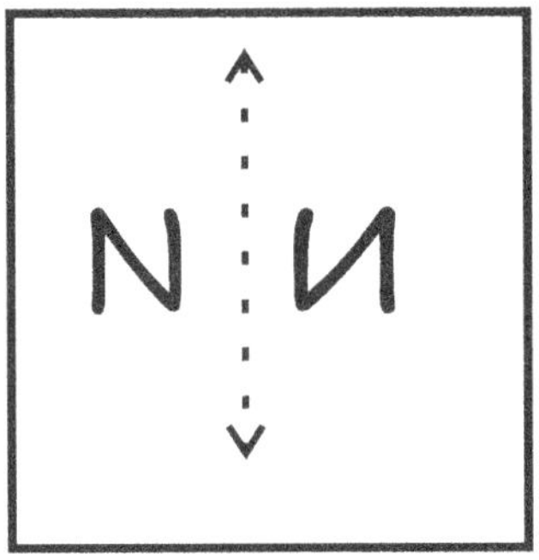

(B)

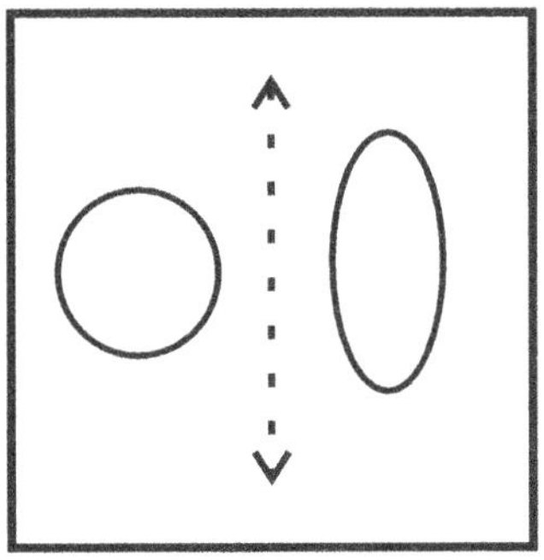

(C)

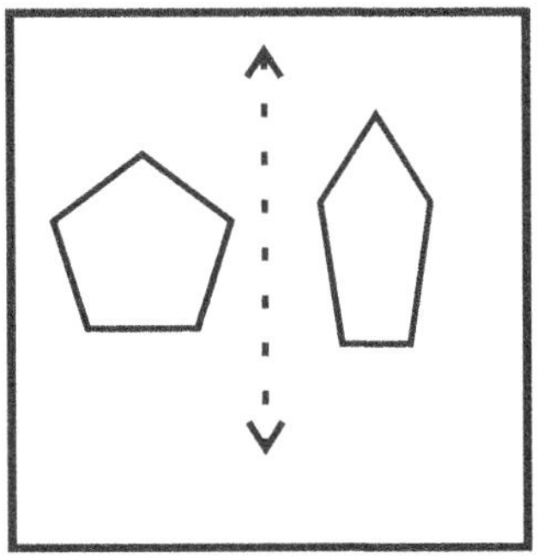

(D)

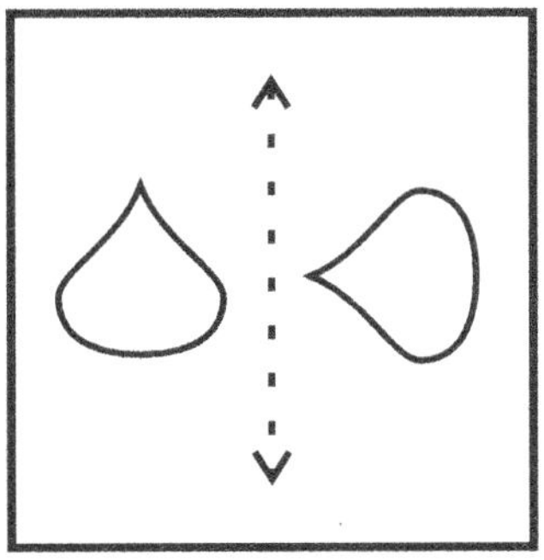

49. Which of the below options represents a translation of four units to left and three units to down ?

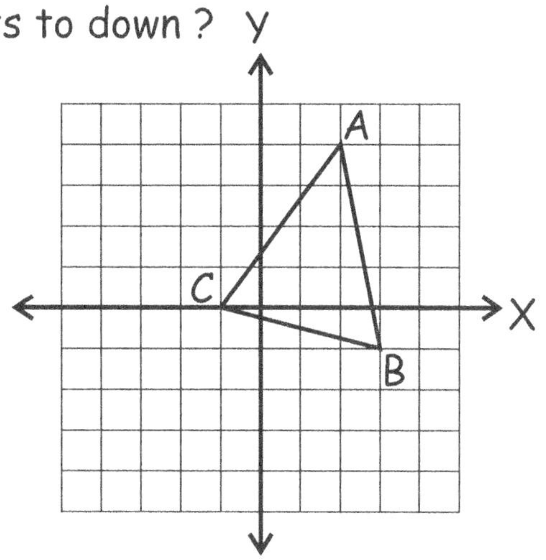

(A)

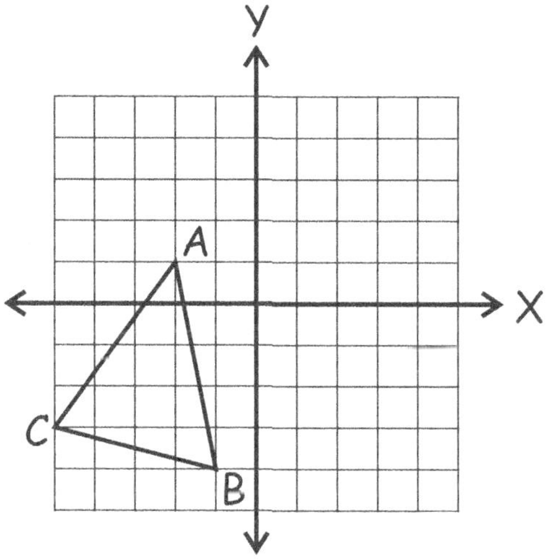

(B)

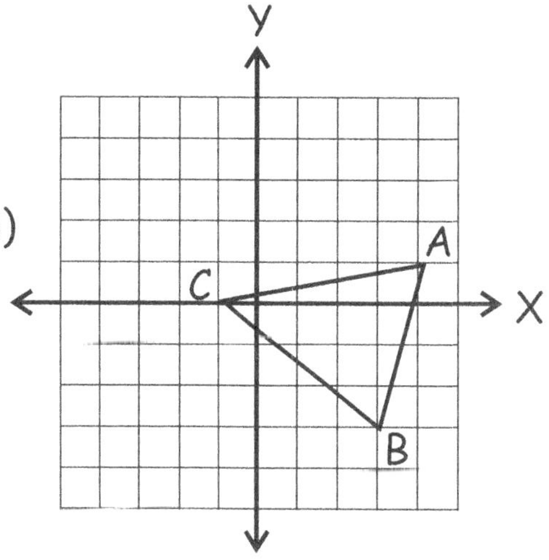

(C)

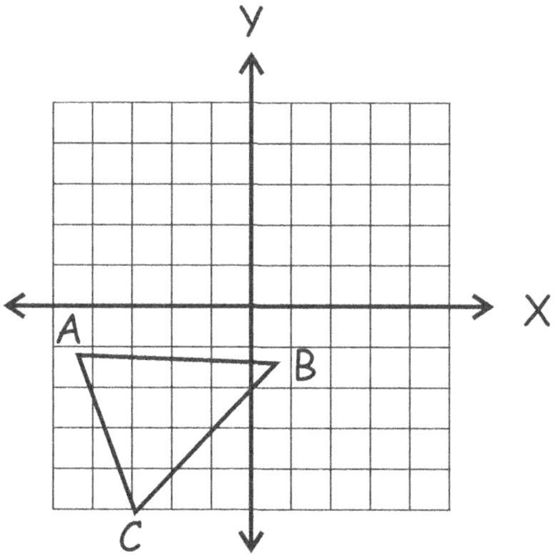

(D)

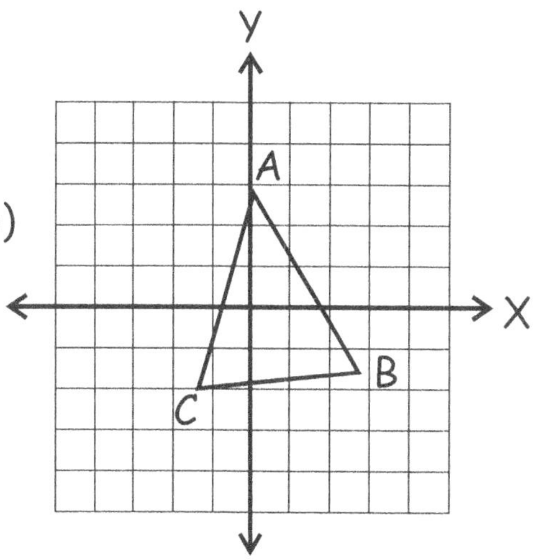

50. Tom wants to order lunch for his team. He must choose 1 pizza, 1 burger and 1 side to order from the below.

Food Combinations

Type of pizza	Type of burger	Side
Pepperoni	Cheese Sausage	Chips Soda

Based on the information in the chart, which tree diagram shows all of Tom's possible combinations ?

(A) Pepperoni pizza ⟶ Cheese burger ⟶ Chips

(B)

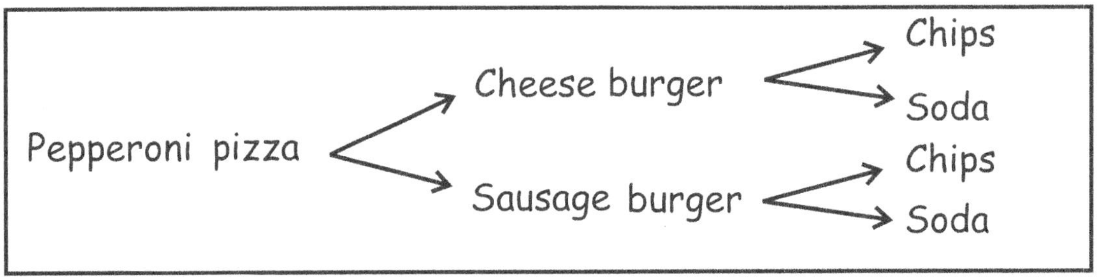

(C)

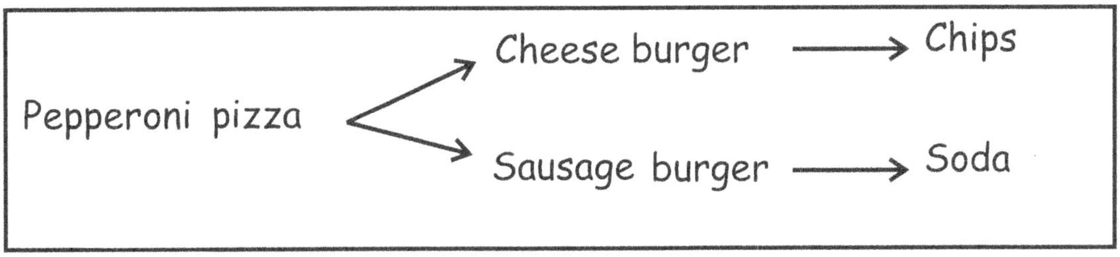

(D)

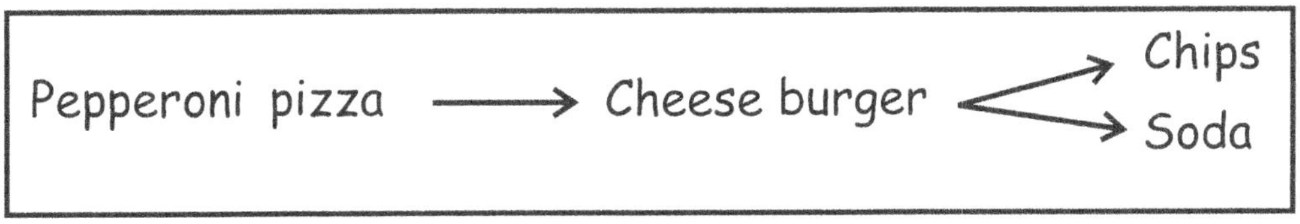

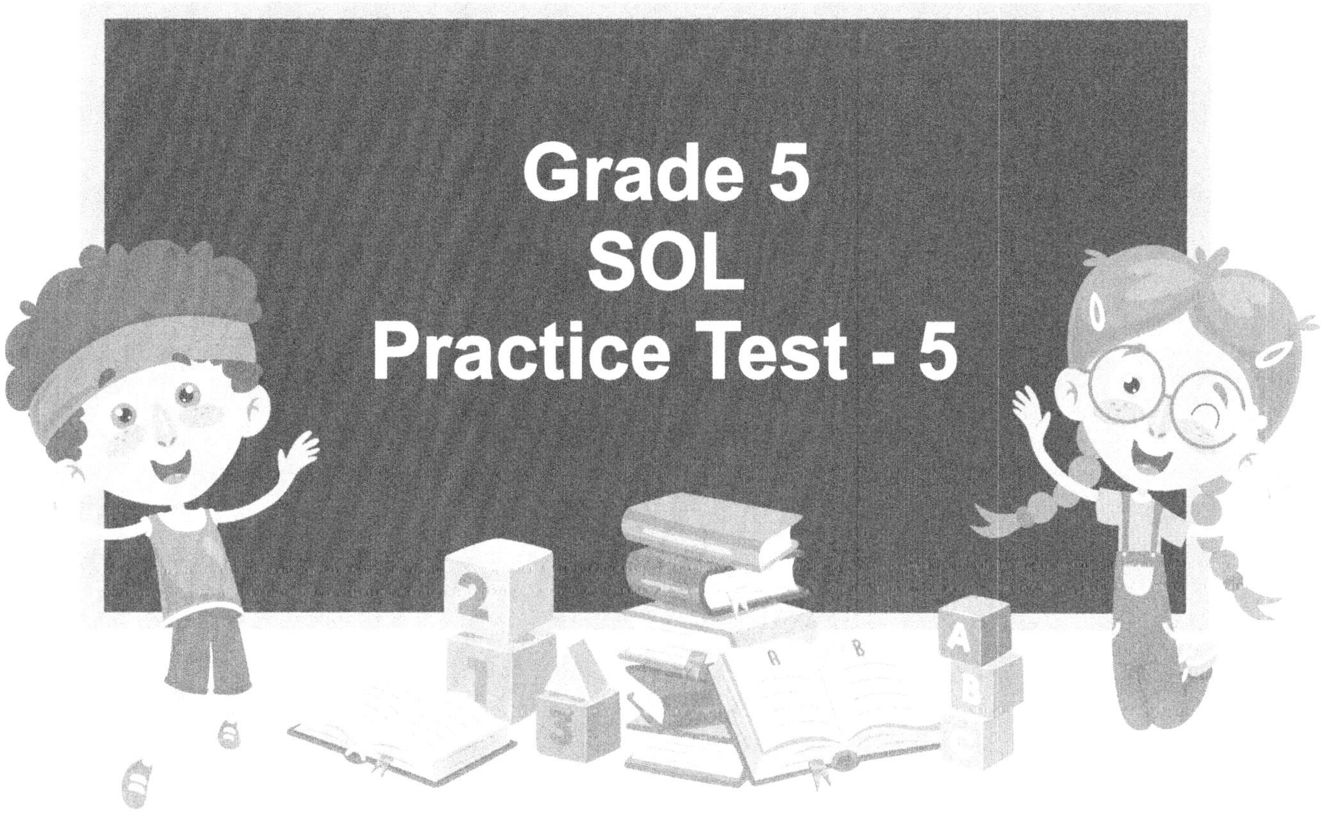
Grade 5
SOL
Practice Test - 5

1. Evaluate $2\frac{3}{4} + 5\frac{1}{4} = \boxed{?}$

(A) 9.65 (B) 8.60 (C) 8.55 (D) 8.0

2. Evaluate $2889 \div 3 = \boxed{?}$

(A) 963 (B) 9653 (C) 961 (D) 962

3. What is the sum of $\frac{3}{11} + \frac{1}{2} = ?$

(A) $1\frac{1}{11}$ (B) $7\frac{1}{22}$

(C) $\frac{17}{22}$ (D) $3\frac{7}{22}$

4. Evaluate

$$\begin{array}{r} 902.01 \\ \times \quad 0.3 \\ \hline \\ \hline \end{array}$$

(A) 277.63 (B) 270.603

(C) 270.633 (D) 270.63

5. Evaluate $10656 \div 12 = \boxed{?}$

(A) 888 R 0

(B) 883 R 7

(C) 888 R 3

(D) 886 R 5

6. Amy bought 9 cases of water bottles. If each case has 18 bottles . How many water bottles did she purchased ?

(A) 207

(B) 96

(C) 165

(D) 162

7. Evaluate

$31.08 \div 4 = \boxed{?}$

(A) 7.17 (B) 77.07 (C) 7.77 (D) 7.077

8. $\frac{2}{11} + \frac{5}{11} = ?$

(A) $\frac{7}{11}$

(B) $\frac{6}{11}$

(C) $\frac{5}{9}$

(D) $\frac{5}{11}$

9. Stacy bakes cupcakes today as per the below table

Baking cup cakes	**Quantity**
Batch 1	18
Batch 2	12
Batch 3	16
Batch 4	17

How many cupcakes did Stacy baked today ?

(A) 64 (B) 66

(C) 65 (D) 63

10. Evaluate

$$\begin{array}{r} 77.07 \\ -\ 17.52 \\ \hline \\ \hline \end{array}$$

(A) 59.59 (B) 55.55

(C) 59.55 (D) 53.35

11. $\frac{7079}{7} = ?$

(A) 1011 R 2 (B) 1011 R 7

(C) 1011 R 0 (D) 1011R 3

12. Ben solved 82 puzzles out of 105 puzzles that he received as his Christmas gift. How many more puzzles does he need to solve ?

(A) 21 (B) 25

(C) 23 (D) 22

13. Order the below fractions from least to greatest ?

$$\frac{5}{8}, \frac{1}{6}, \frac{7}{8}, \frac{3}{4}$$

(A) $\frac{5}{8}, \frac{1}{6}, \frac{7}{8}, \frac{3}{4}$ (B) $\frac{3}{4}, \frac{5}{8}, \frac{1}{6}, \frac{7}{8}$

(C) $\frac{3}{4}, \frac{7}{8}, \frac{1}{6}, \frac{5}{8}$ (D) $\frac{1}{6}, \frac{5}{8}, \frac{3}{4}, \frac{7}{8}$

14. What is the value of 8 in 7700.9851 ?

(A) Eight thousandth (B) Eight

(C) Eight hundredth (D) Eight tenth

15. Which of the below statement is correct ?

(A) 44.042 < 42.502 (B) 44.004 > 44.402

(C) 44.99 > 44.45 (D) 44.04 > 44.401

16. Round 999.802 to the nearest thousandths

(A) 999.800 (B) 999.903

(C) 999.802 (D) 999.850

17. 807.253 written or read in words as ?

(A) Eight hundred and two hundred fifty-three hundreds

(B) Eight hundred and two hundred fifty-three thousands

(C) Eight hundred seven and two hundred fifty-three hundredths

(D) Eight hundred seven and two hundred fifty-three thousandths

18. Evaluate $\frac{555}{1000} = \boxed{?}$

(A) 5.56 (B) 0.555 (C) 0.0555 (D) 5.556

19. 37.5486 round to near thousandths ?

(A) 37.55 (B) 37.6. (C) 37.549 (D) 37.55

20. Which of the below shows a straight angle ?

(A)
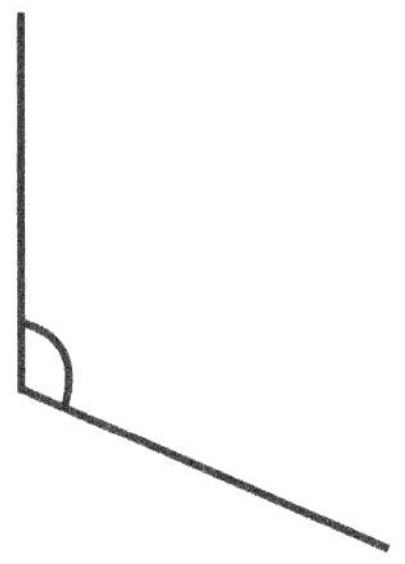

(B)

(C)
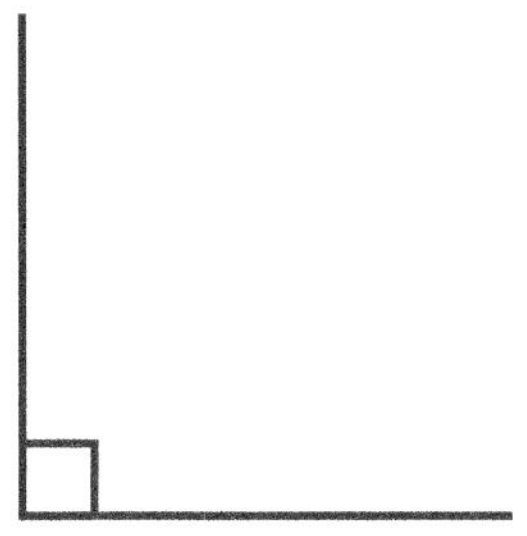

(D)

21. Which of the below figures has six sides ?

(A) 

(B) 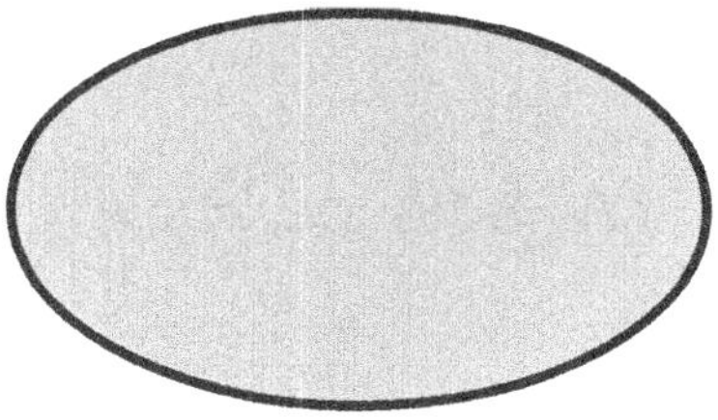

(C) 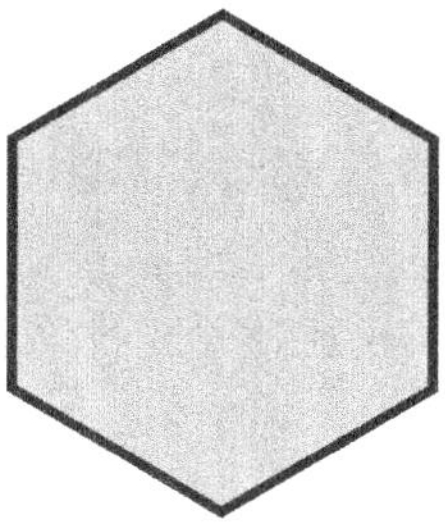

(D) 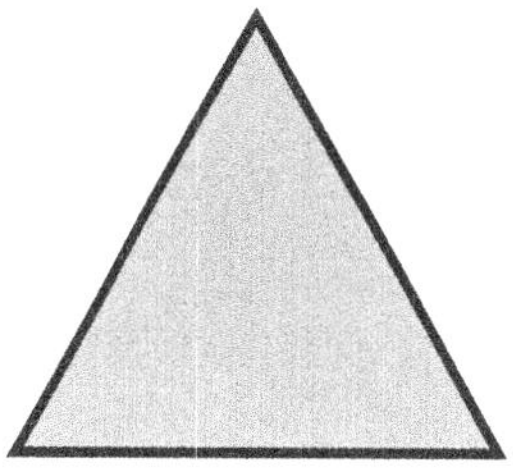

22. What types of angles are used to form the below triangle ?

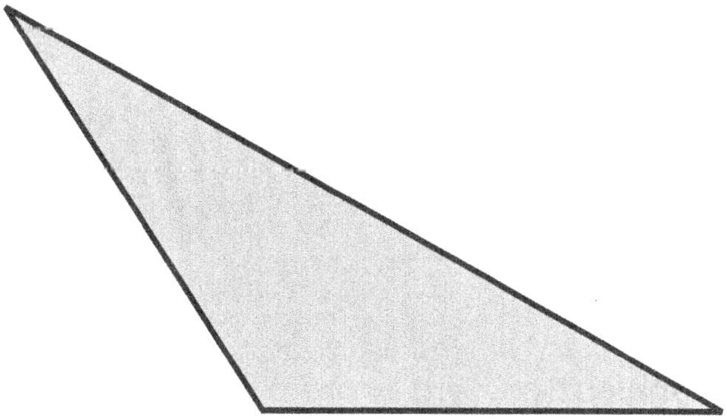

(A) Acute angle , Obtuse angle , Acute angle
(B) Acute angle , Right angle , Acute angle
(C) Acute angle , Acute angle , Acute angle
(D) Acute angle , Straight angle , Obtuse angle

23. Rita is reading a map in during her expedition trip in the forest. What does she need to find in which direction she needs to go ?

(A) Balance Scale (B) Thread

(C) Ruler (D) Compass

24. Phillip has a vegetable garden in the form of hexagon with equal sides of length 17 meters. what is the perimeter of the vegetable garden ?

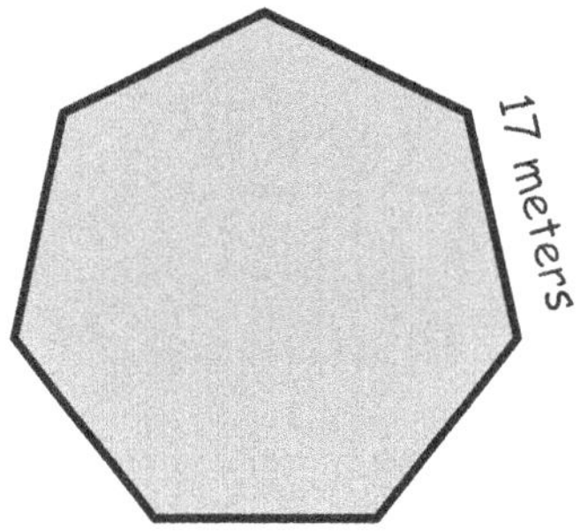

(A) 60 meters (B) 102 meters

(C) 169 sq.meters (D) 78 meter

25. Andy does his work out at gym from 6 AM to 7.30 AM from Monday through Friday. How many hours did Andy workout at gym this week ?

(A) 6 hours 35 minutes
(B) 7 hours 20minutes
(C) 6 hours 30 minutes
(D) 7 hours 30 minutes

26. Find the area of the square plot below with the side length of 20 miles ?

(A) 400 sq.miles (B) 160 sq.miles

(C) 80 miles (D) 400 miles

27. Mrs. Sandy wants to buy an empty land to construct a pre-school. What does she need to know before buying the piece of land ?

(A) Width (B) Volume

(C) Area (D) Length

28. Milli's study room has an area of 120 sq centimeters and perimeter of 46 centimeters. What are dimensions of the room ?

(A) Length — 14 centimeters ; width — 5 centimeters
(B) Length — 25 centimeters ; width — 3 centimeters
(C) Length — 30 centimeters ; width — 4 centimeters
(D) Length — 15 centimeters ; width — 8 centimeters

29. O is the center of the circle. The line p touches the circle at the point B. What is the line called ?

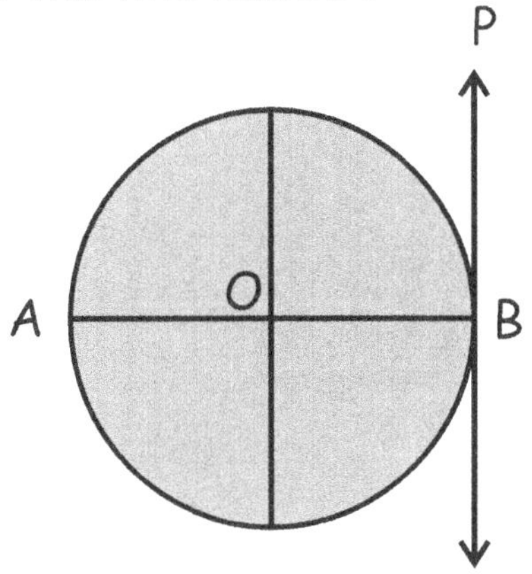

(A) Diameter (B) Tangent

(C) Chord (D) Arc

30. Which of the below picture has one vertex , one edge and one flat face ?

(A)
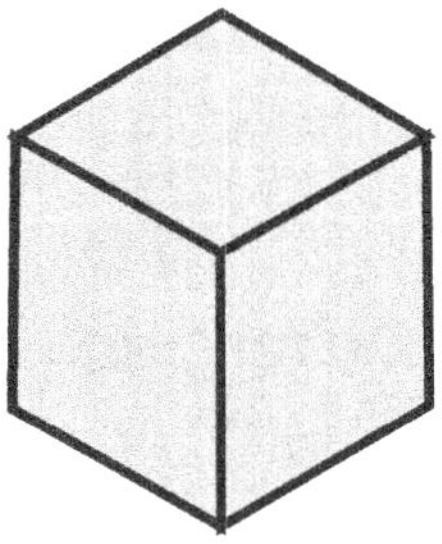

(B)
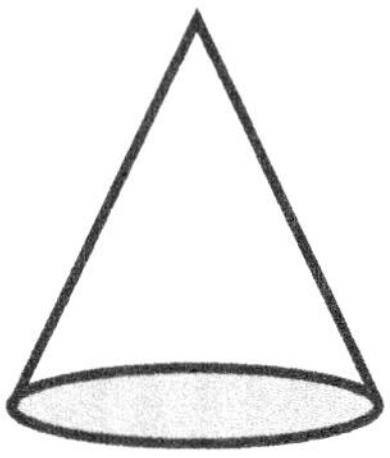

(C)
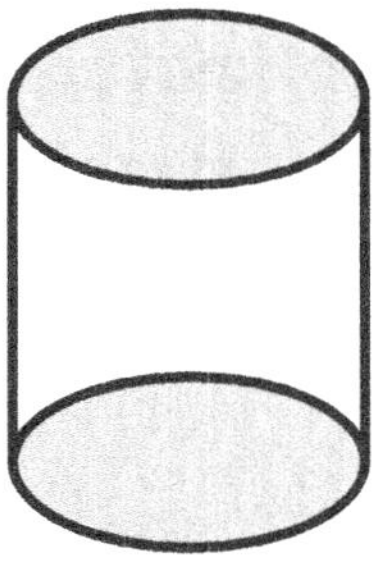

(D)
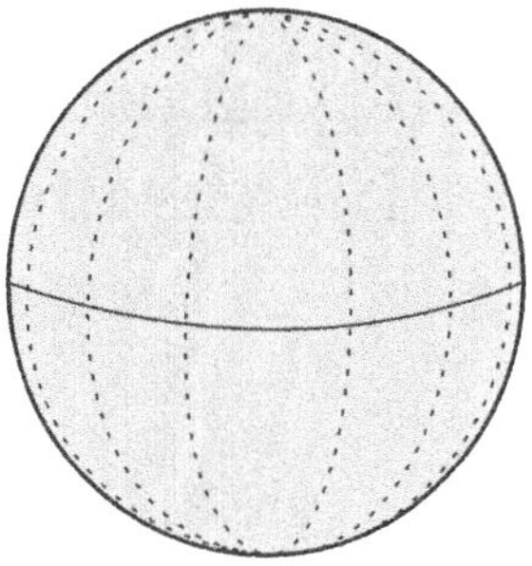

31. Calculate the average of the below data set ?

5 , 1 , 7 , 9 , 11 , 6 , 5 , 2 , 9 , 5

(A) 12 (B) 10 (C) 5 (D) 6

32. Weight of a feather will be ?

(A) < Pounds (B) > Ounces

(C) > Pound (D) < Ounces

33. Which of the below picture shows the radius AB ?

(A)

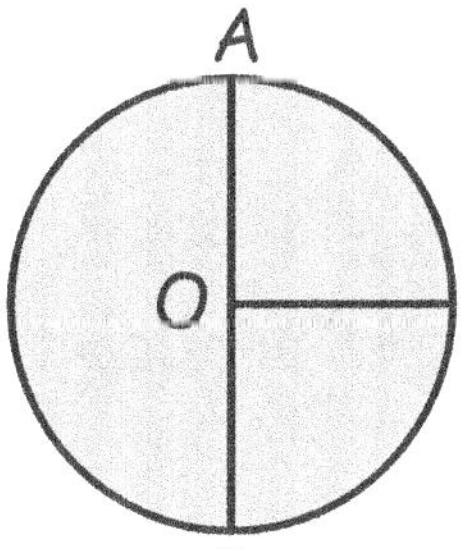

(B)

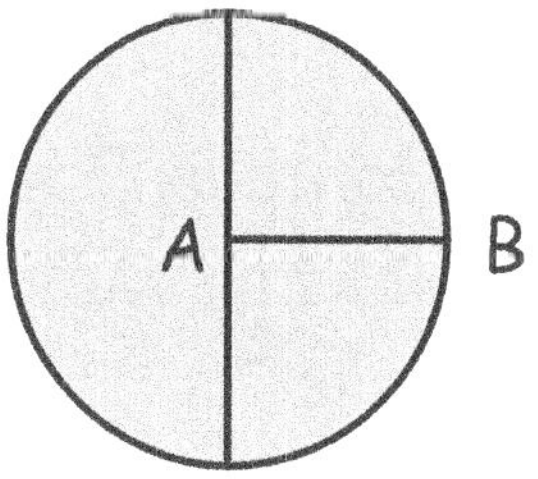

(C)

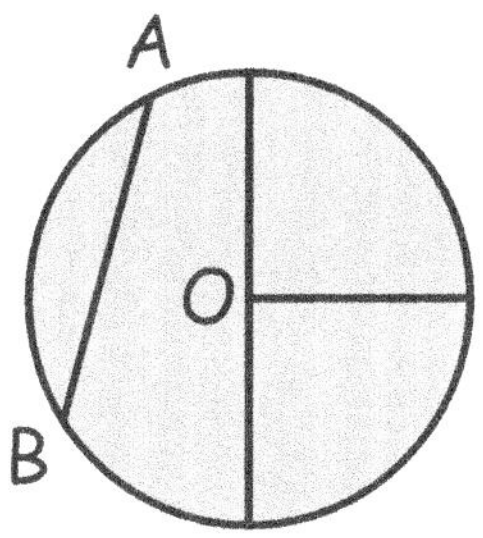

(D)

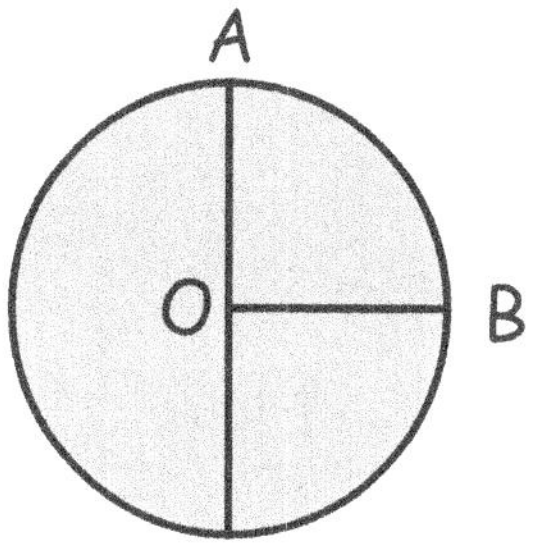

34. Pam makes a spin wheel to choose a car color to purchase. What is the probability of choosing a Black car in one spin ?

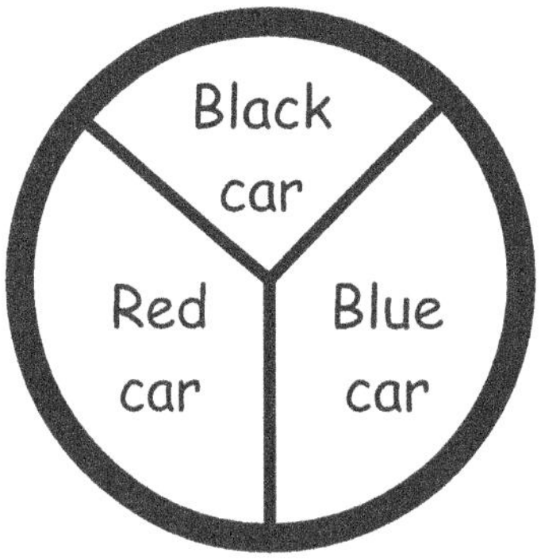

(A) $\frac{1}{8}$ (B) $\frac{1}{6}$

(C) $\frac{1}{3}$ (D) $\frac{2}{5}$

35. A box contains 9 white clips , 11 Black clips and 8 Gold clips. Zara picks up one clip from the box. What is the probability of selecting a Gold clip ?

(A) $\frac{2}{7}$ (B) $\frac{11}{28}$

(C) $\frac{5}{7}$ (D) $\frac{1}{8}$

36. Julian made the following list of all his math quiz scores.

11 , 27 , 15 , 29 , 28 , 29 , 35 , 39 , 20 , 31

Which stem-and-leaf plot correctly shows Julian's quiz scores?

(A)

Stem	Leaf
1	0 , 5
2	0 , 7 , 1 , 8 , 9
3	1 , 8 , 9

(B)

Stem	Leaf
1	1 , 8
2	1 , 5 , 7 , 8 , 9
3	0 , 5 , 7

(C)

Stem	Leaf
1	1 , 5
2	5 , 7 , 8 , 8 , 9
3	1 , 5 , 9

(D)

Stem	Leaf
1	1 , 5
2	0 , 7 , 8 , 9 , 9
3	1 , 5 , 9

37. Gia and Sia are choosing the menu to order for their dinner.

Spinner A

Spinner B

If Jia chooses Pizza on the Spinner A. Show all possible outcomes in Spinner B in the form of a tree diagram.

(A) Pizza → Bread Sticks
Pizza → Salad
Pizza → Cheese Knots

(B) Pizza → Bread Sticks
Pizza → Pasta
Pizza → Salad

(C) Pizza → Sandwich
Pizza → Pasta
Pizza → Cheese Knots

(D) Pizza → Bread Sticks
Pizza → Sandwich
Pizza → Cheese Knots

38. The bouncy boys soccer team played various games this quarter. The below graph shows the number of goals made by the team in each game.

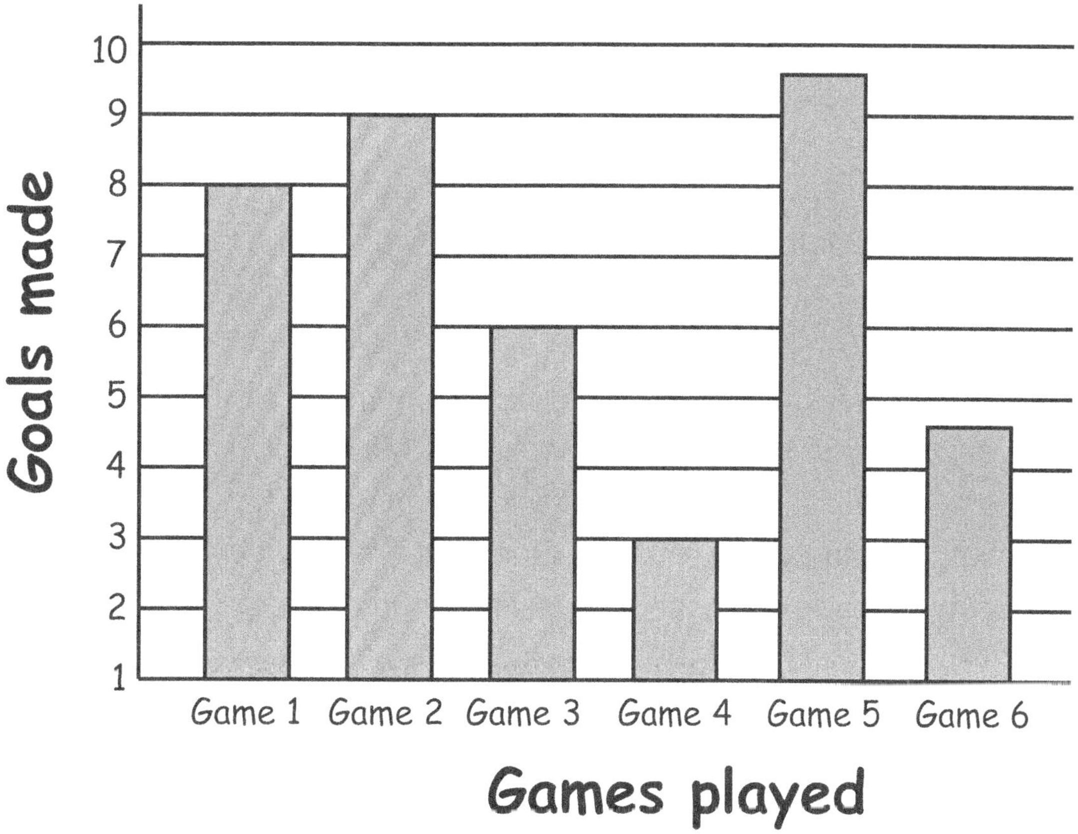

The greatest difference in the number of goals made between the

(A) Game 1 and Game 2

(B) Game 3 and Game 6

(C) Game 4 and Game 5

(D) Game 4 and Game 6

39. Find the mode for the following data set of numbers ?

105 , 109 , 111 , 117 , 117 , 119 , 109 , 109 , 117 , 117

(A) 117
(B) 107
(C) 109
(D) 111

40. Translate the below open sentence into words.

$$2h - 1 = 9$$

(A) Angela sold h cupcakes. Twice the number of cupcakes and one equals to 9. How many cupcakes did she sell ?

(B) Angela baked h cupcakes. One less than twice the number of cupcakes baked equals to 9. How many cupcakes did she bake ?

(C) Angela baked h cupcakes. The number of cupcakes baked equals to 9

(D) Angela sold h cupcakes. Twice the number of cupcakes equals to 9

41. What is the variable in the expression $\frac{9y}{5}$?

(A) 5
(B) 9
(C) X
(D) Y

42. Which of the below options represents a translation of three units to the right ?

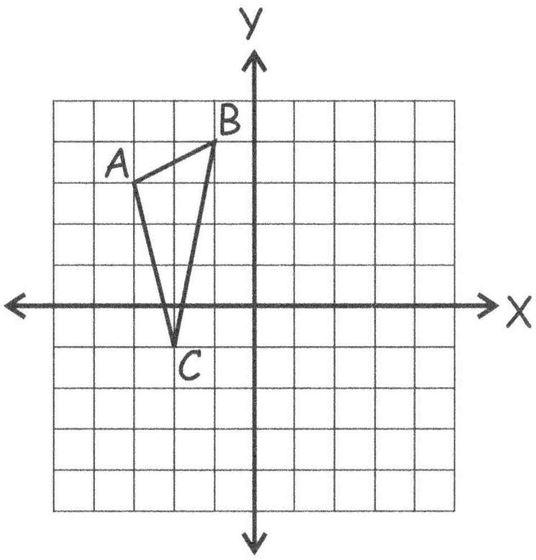

(A)

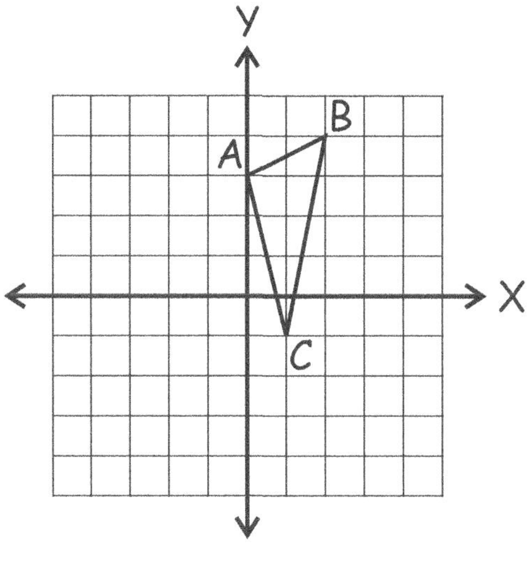

(B)

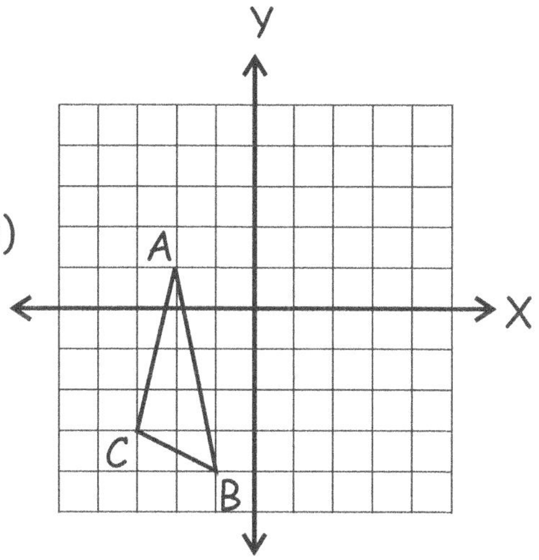

(C)

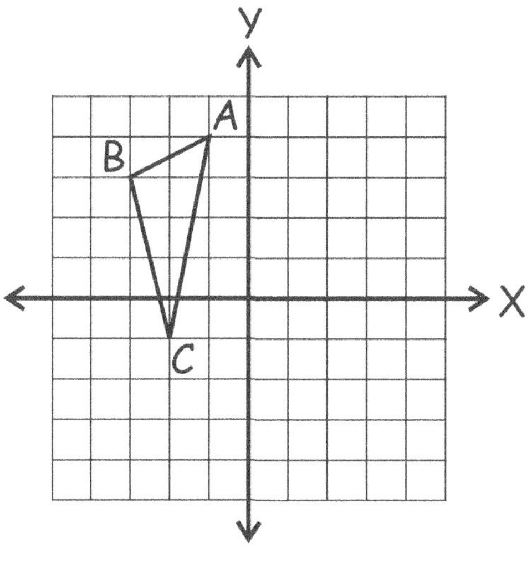

(D)

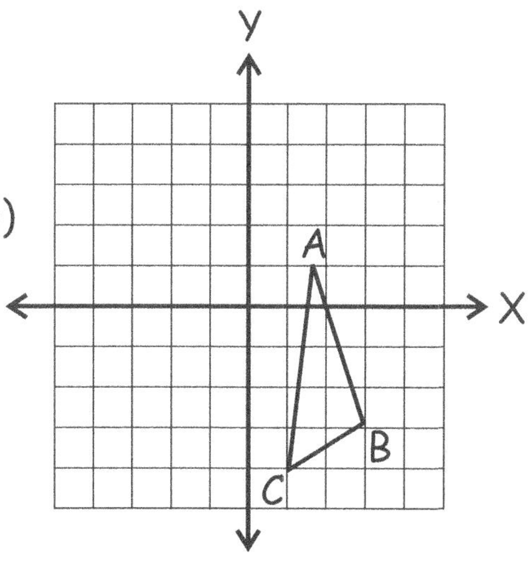

43. Kelly is making bracelets and is planning to sell them at $4 per piece. If "b "is represents bracelets which of the below expression displays the money she can earn ?

(A) $b - 4$ (B) $b + 4$

(C) $\frac{b}{4}$ (D) $4 * b$

44. A stationary store sells "p" pencils in one box. Tracy bought 7 boxes for $m. What is the price of one box ?

(A) $\frac{m}{7}$ (B) $m \times 7$

(C) $\frac{7}{m}$ (D) $7 + p$

45. Henkel distributes 1730 apples among 12 hospitals, equally representing the equation $1730 \div 12 = 144$ R p. What does p represent in the equation ?

(A) Number of apples each person received

(B) Number of hospitals

(C) Number of apples left

(D) Number of apples each hospitals received

46. If "b" books are distributed among 24 kids in the class. Which of the below expression represents the number of books each student receives ?

(A) $\frac{b}{24}$ (B) b*24.

(C) b+24 (D) b - 24

47. Number sentence 7h - 1 = 13 can be written in words as

(A) Bella baked h muffins in 7 batches and ate one of them . 13 muffins are left out. How many muffins did she bake all together ?

(B) Bella baked 7 muffins in h batches leaving 13 muffins. How many muffins did she bake in each batch ?

(C) Bella baked h muffins in 1 batches and ate 7 of them. 13 muffins are left. How many muffins did she bake in total ?

(D) Bella baked 3 muffins in 2 batches . How many muffins does she need to bake more ?

48. Tom does twice as many pushups as his friend Matt. If Matt does 12 pushups. Which number sentence could be used to find *p*, the number of pushups that tom did ?

(A) 2 ÷ 12
(B) p = 12 - 2
(C) p = 12 * 2
(D) m = 12 + 2

49. This table below shows the track practice time by crazy runners team this week.

Day	Time (hr)
1	1.22
2	1.35
3	1.48
4	2.01

If the pattern continues, What is the time spent by super team at the end of 5th day ?

(A) 2 hours , 24 minutes

(B) 2 hours , 14 minutes

(C) 1 hours , 94 minutes

(D) 2 hours , 28 minutes

50. Which of the below pair of pictures are symmetric to each other ?

(A)

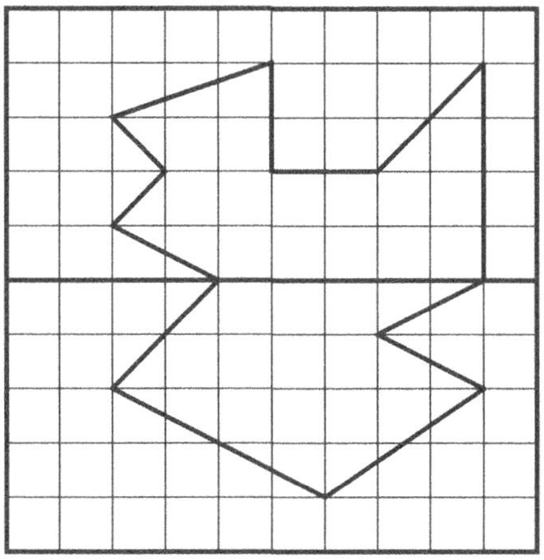

(B)

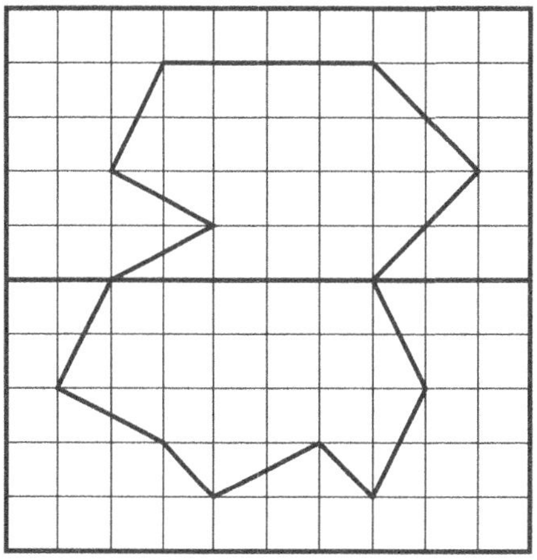

(C)

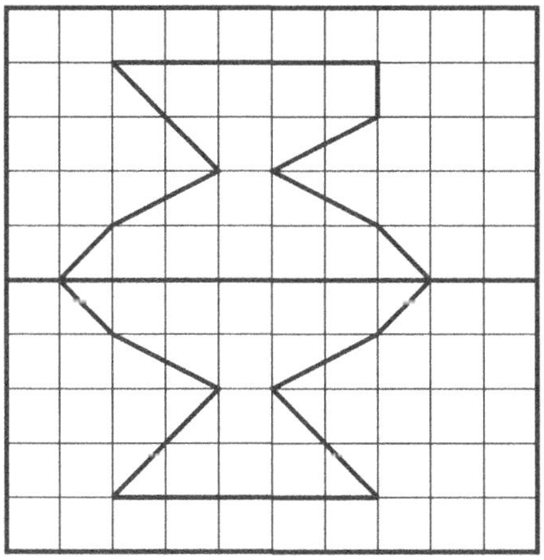

(D)

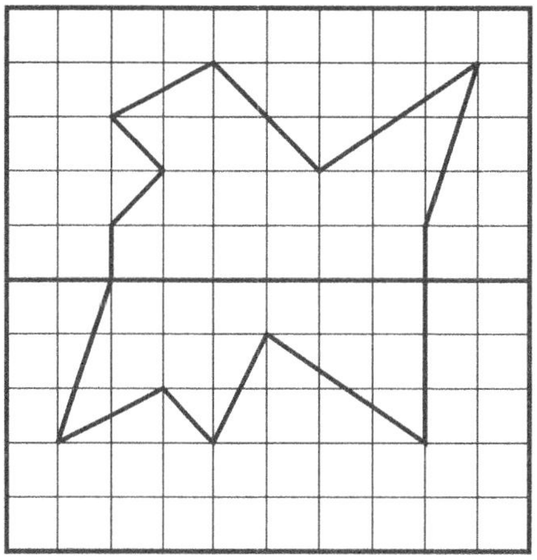

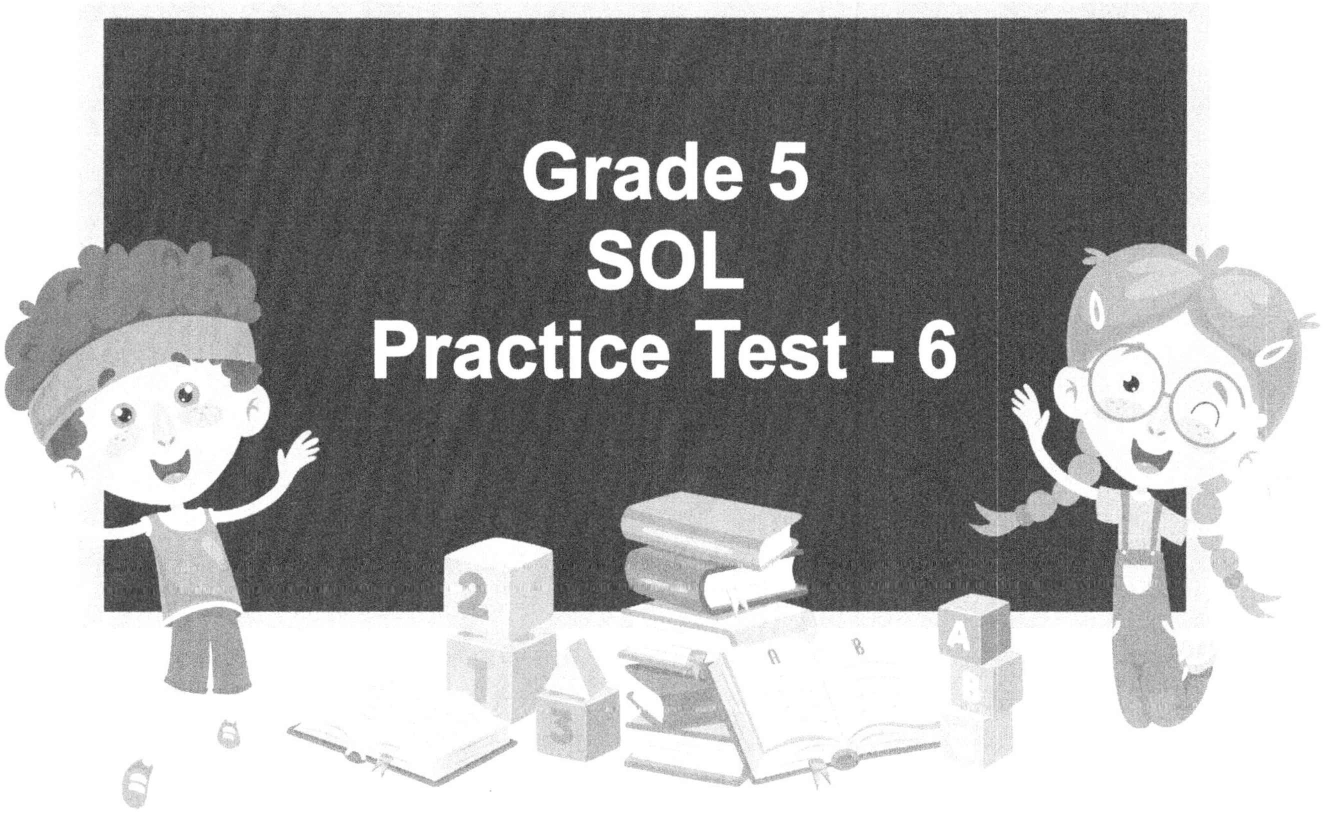
Grade 5
SOL
Practice Test - 6

1. $7\frac{3}{5} - 3\frac{1}{2} = ?$

(A) 4.01 (B) 4.67 (C) 4.1 (D) 11.1

2. Evaluate $18432 \div 6 = \boxed{?}$

(A) 3070 R 5
(B) 6144 R 1
(C) 3802 R 0
(D) 3072 R 0

3. What is he sum of $\frac{5}{7} + \frac{3}{7} = ?$

(A) $1\frac{1}{7}$
(B) $1\frac{2}{7}$
(C) $\frac{1}{7}$
(D) $3\frac{2}{7}$

4. Evaluate

$$\begin{array}{r} 25.31 \\ \times \quad 1.1 \\ \hline \\ \hline \end{array}$$

(A) 26.841
(B) 27.831
(C) 27.841
(D) 28.741

5. Evaluate $467 \div 22 = \boxed{?}$

(A) 20 R 21 (B) 21 R 5

(C) 21 R 6 (D) 20 R 11

6. Tom bought 3 boxes of Ping-Pong balls. If each box has 24 Ping-Pong balls. How many Ping-Pong balls did he purchased ?

(A) 78 (B) 27

(C) 48 (D) 72

7. Evaluate

$15.75 \div 5 = \boxed{?}$

(A) 3.15 (B) 3.05 (C) 3.25 (D) 3.015

8. Evaluate $\frac{5}{8} + \frac{1}{4} = ?$

(A) $\frac{7}{8}$ (B) $\frac{1}{8}$

(C) $\frac{3}{4}$ (D) $\frac{6}{8}$

9. Ryan played in a basketball tournament this weekend. The number of baskets made by him are shown below.

Games (Saturday Dec 22 and Sunday Dec 23)	No of Baskets
Dec 22-Game1: 9 AM-10.30 AM	12
Dec 22-Game2: 3 PM-4.30 PM	19
Dec 23-Game1: 11 AM-12.30 PM	13
Dec 23-Game2: 3 PM-4.30 PM	13

How many baskets did Ryan made this weekend ?

(A) 56 (B) 55 (C) 65 (D) 57

10. Evaluate

```
   99.08
 - 37.19
-----------------

---------------
```

(A) 61.88 (B) 61.89

(C) 60.89 (D) 62.89

11. Evaluate $\frac{4289}{21} = ?$

(A) 204 R 5 (B) 24 R 5

(C) 214 R 6 (D) 204 R 3

12. Jack sold 523 tickets for school basketball game out of 767 tickets. How many more tickets does he need to sell ?

(A) 201 (B) 204

(C) 244 (D) 264

13. Order the below fractions from least to greatest ?

$\frac{7}{9}, \frac{2}{3}, \frac{2}{6}, \frac{1}{9}, \frac{4}{9}$

(A) $\frac{7}{9}, \frac{2}{3}, \frac{2}{6}, \frac{1}{9}, \frac{4}{9}$ (B) $\frac{2}{6}, \frac{4}{9}, \frac{1}{9}, \frac{2}{3}, \frac{7}{9}$

(C) $\frac{2}{3}, \frac{7}{9}, \frac{1}{9}, \frac{2}{6}, \frac{4}{9}$ (D) $\frac{1}{9}, \frac{2}{6}, \frac{4}{9}, \frac{2}{3}, \frac{7}{9}$

14. What is the value of 5 in 7.453 ?

(A) Five thousandth
(B) Five
(C) Five hundredth
(D) Five tenth

15. Which of the below statement is correct ?

(A) 77.47 < 77.407
(B) 77.502 > 77.512
(C) 77.506 < 77.510
(D) 77.889 > 77.899

16. Round 76.476 to the nearest tenths

(A) 76.500
(B) 77.00
(C) 76.48
(D) 76.47

17. 37.421 written or read in words as ?

(A) Thirty-seven and four hundred two-one hundreds

(B) Thirty-seven and four hundred two-three thousands

(C) Thirty-seven and four thundred two-one hundredths

(D) Thirty-seven and four hundred twenty-one thousandths

18. Evaluate $\frac{31}{1000}$ = ?

(A) 0.31 (B) 0.031 (C) 0.3 (D) 0.0031

19. Round 3124.561 to near hundredths ?

(A) 3124.56 (B) 3124.500

(C) 3125 (D) 3124.600

20. Which of the below shows a right angle ?

(A)

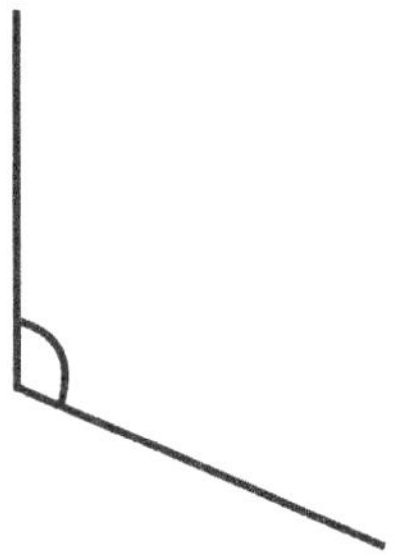

(B)

(C)

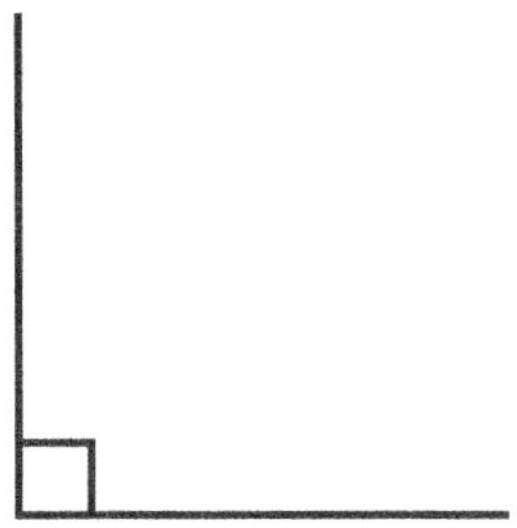

(D)

21. Which of the below figures has eight sides ?

(A) 

(B) 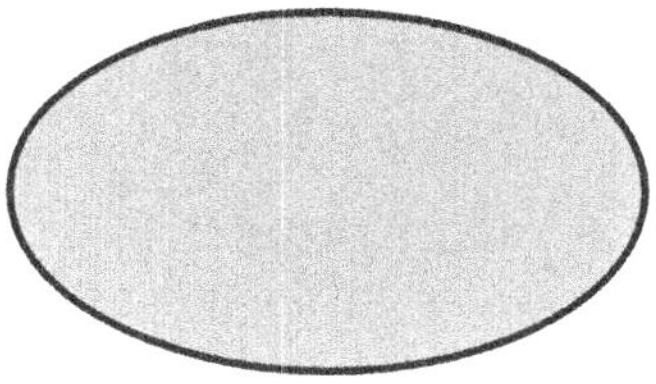

(C) 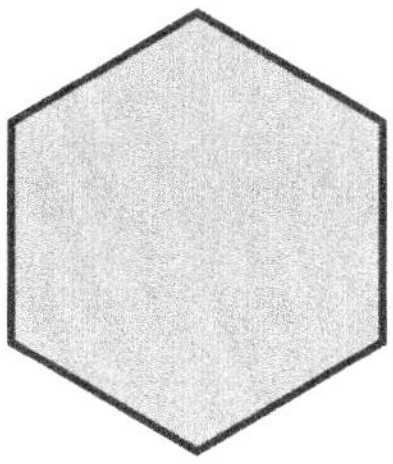

(D) 

22. What types of angles are used to form the below triangle ?

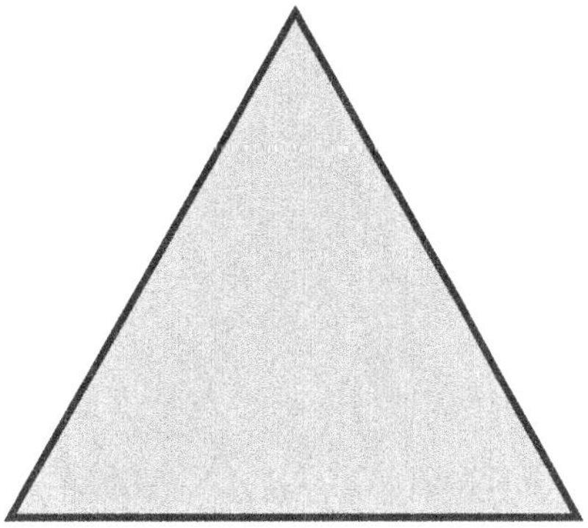

(A) Acute angle , Obtuse angle , Acute angle
(B) Acute angle , Acute angle , Acute angle
(C) Acute angle , Right angle , Acute angle
(D) Acute angle , Straight angle , Obtuse angle

23. Rosy the architect draws the shopping complex, Garden around it and the parking spaces. After drawing she wants to measure the length and width of the parking lot. She needs which of the below to measure the distance ?

(A) Compass

(B) Thread

(C) Ruler

(D) Balance Scale

24. Phillip has a vegetable garden in the form of octagon with equal sides of length 10 meters. what is the perimeter of the vegetable garden ?

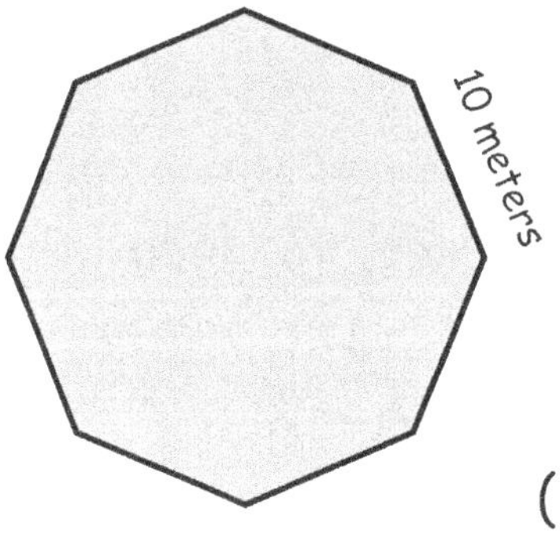

(A) 80 meters

(B) 102 meters

(C) 169 sq.meters

(D) 78 meter

25. Bryan starts mowing the lawn at 8:15 AM. He finished the lawn mowing at 4:26 PM. If he takes lunch break of 2 hours in between, How long did bryan took to mow the lawn ?

(A) 3 hours 21 minutes

(B) 6 hours 31 minutes

(C) 5 hours 11 minutes

(D) 6 hours 11 minutes

26. Find the area of the square plot below with the side length of 13 kilometers ?

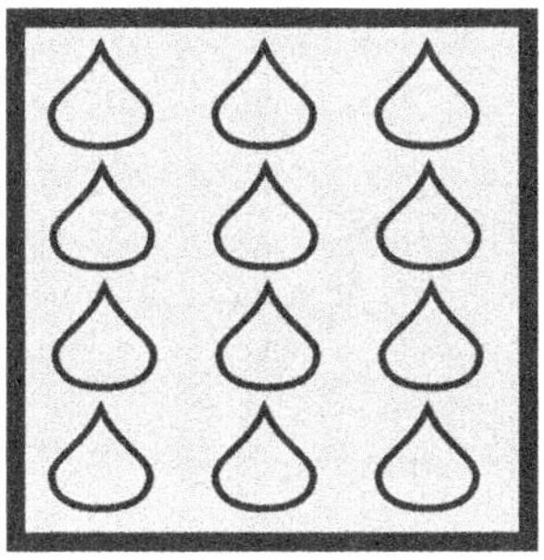

(A) 52 sq.km (B) 169 sq.km

(C) 149 km (D) 26 sq.km

27. Ms. Dona fixes a fence around her garden. Total length of the fence is an example of

(A) Width (B) Area

(C) Volume (D) Perimeter

28. Lilys study room has an area of 24 sq centimeters and 28 centimeters ?

(A) Length — 1 inch; width — 24 inches
(B) Length — 2 inches; width — 14 inches
(C) Length — 2 inches; width — 20 inches
(D) Length — 2 inches; width — 12 inches

29. O is the center of the circle. then AB is called ?

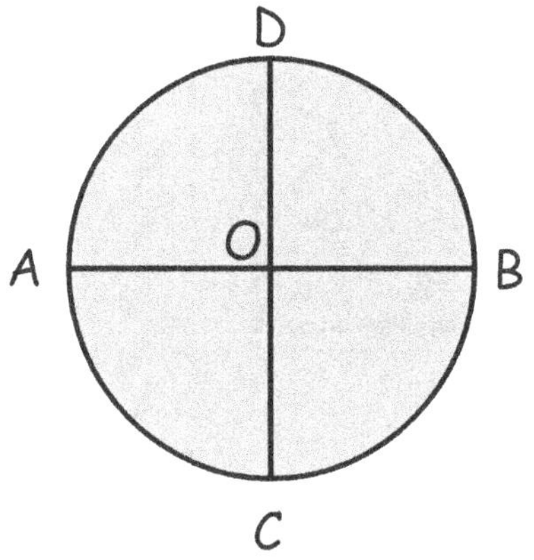

(A) Diameter (B) Radius

(C) Chord (D) Arc

30. Which of the below picture has eight vertices and twelve edges ?

(A) 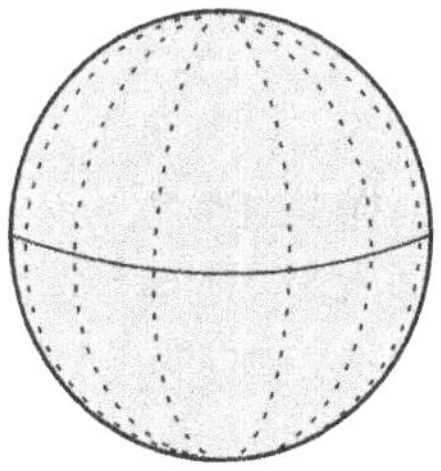

(B) 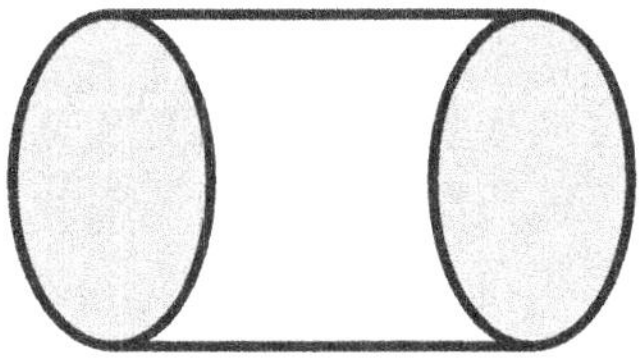

(C) 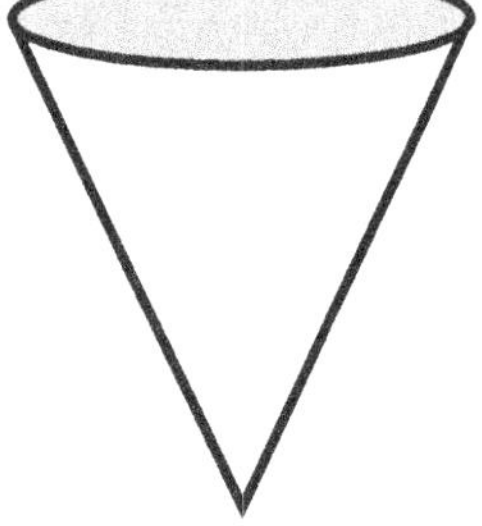

(D) 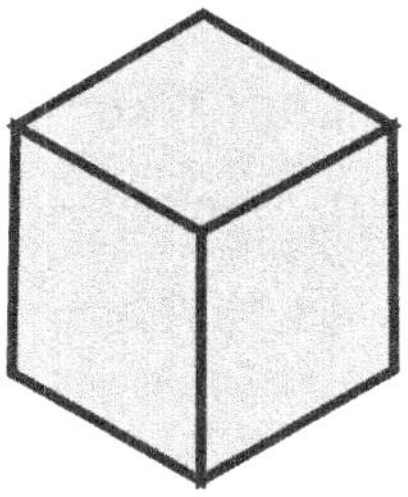

31. Calculate the average of the below data set ?

11 , 11 , 15 , 14 , 19 , 20

(A) 16 (B) 14 (C) 15 (D) 12

32. Amount of juice in one glass is ?

(A) 1 Gallon (B) 2 Quart

(C) 1 Litre (D) 5 Pounds

33. Which of the below picture shows the chord AB ?

(A)
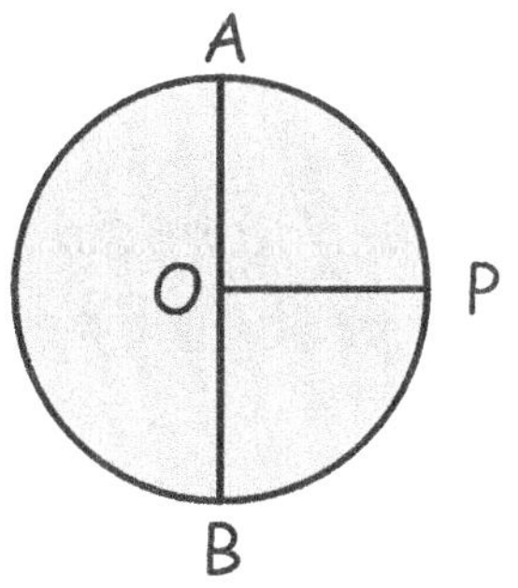

(B)
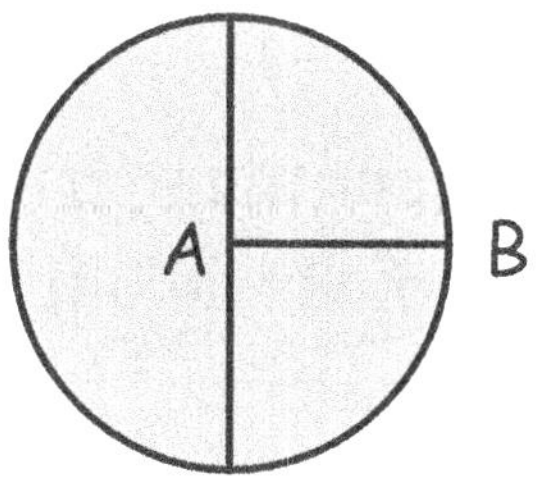

(C)
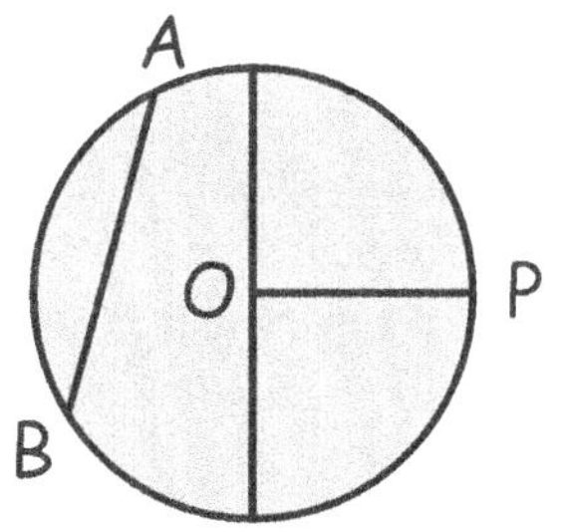

(D)
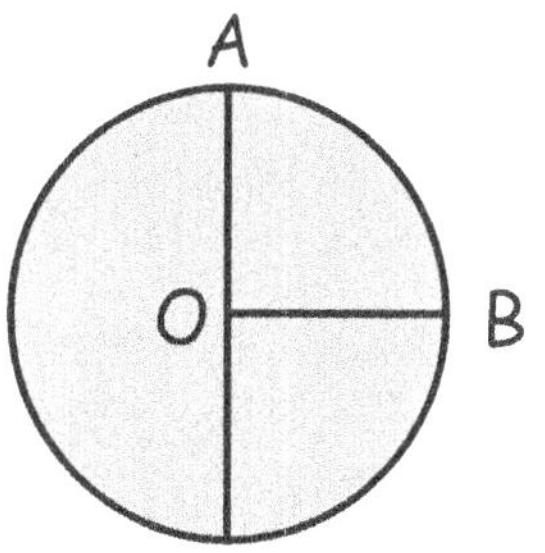

34. Pam plays a spin wheel to win a prize in the fair.
What is the probability of getting a 90 point score ?

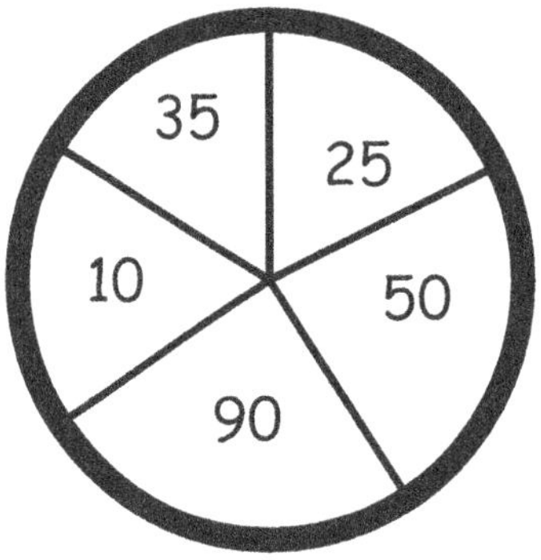

(A) $\frac{1}{5}$ (B) $\frac{3}{4}$

(C) $\frac{1}{8}$ (D) $\frac{3}{5}$

35. A box contains 5 Jelly beans, 3 Candies, and 7 chocolets .
Cathy picks up one from the box. What is the probability of selecting a jelly bean ?

(A) $\frac{7}{15}$ (B) $\frac{3}{15}$

(C) $\frac{1}{5}$ (D) $\frac{1}{3}$

36. Julian made the following list of all his math quiz scores.

78, 79, 76, 76, 86, 77, 90, 81 , 82, 83, 97, 99, 99

Which stem-and-leaf plot correctly shows Julian's quiz scores?

(A)

Stem	Leaf
7	7
8	5 , 6 , 6 , 8 , 8 , 9 , 9
9	1 , 3 , 4 , 9 , 9

(B)

Stem	Leaf
7	7
8	5 , 6 , 6 , 8 , 8 , 9 , 9
9	1 , 3 , 4 , 9 , 9

(C)

Stem	Leaf
7	6 , 6 , 7 , 8 , 9
8	1 , 2 , 3 , 6
9	0 , 7 , 9 , 9

(D)

Stem	Leaf
7	7
8	5 , 6 , 6 , 8 , 8 , 9 , 9
9	1 , 3 , 4 , 9 , 9

37. Jack and Jill are playing a spinning game to choose a snack for them in the evening.

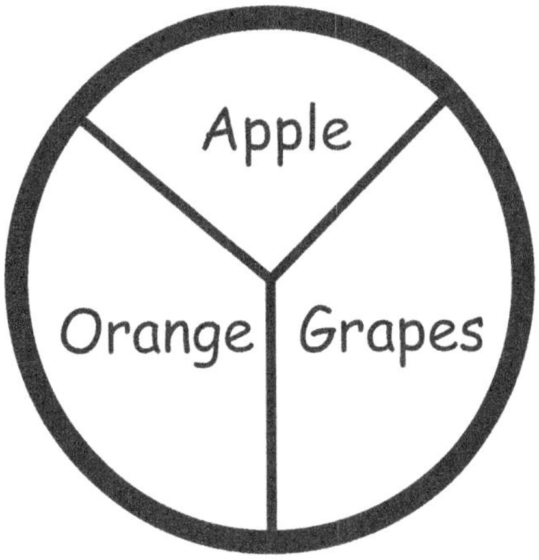

Spinner A

Spinner B

If Jill chooses Grapes on the Spinner A. Show all possible outcomes in Spinner B in the form of a tree diagram.

(A) Grapes → Cake
Grapes → Brownie
Grapes → Cookie

(B) Grapes → Apple
Grapes → Brownie
Grapes → Cookie

(C) Grapes → Cake
Grapes → Orange
Grapes → Cookie

(D) Grapes → Orange
Grapes → Brownie
Grapes → Apple

38. Lucy and her sister are practicing for a marathon. The number of miles they ran together in six weeks of practice are shown below.

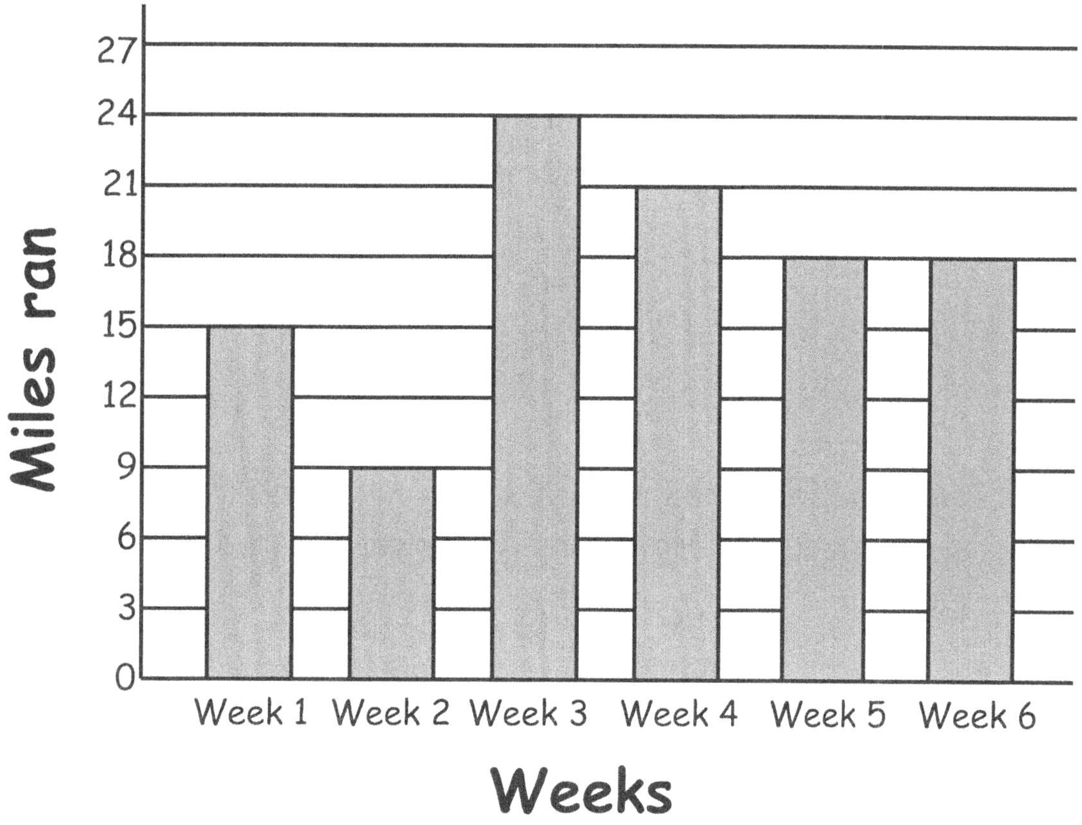

The greatest difference in the number of miles ran is between the

(A) Week 2 and Week 3

(B) Week 1 and Week 6

(C) Week 3 and Week 5

(D) Week 1 and Week 5

39. Find the mode for the following data set of numbers ?

11 , 21 , 23 , 29 , 29 , 31 , 37 , 29 , 39 , 29 , 29

(A) 31 (B) 29

(C) 37 (D) 21

40. Translate the below open sentence into words.

$11 - 5 = X$

(A) Lupin had 11 times more X-box games than his brother who has 5 X-box games. How many X-box games does they have all together now ?

(B) Lupin had 11 X-box games. He bought 5 more. How many X-box games does he have now ?

(C) Lupin had 11 X-box games. His brother has 5 times more than him. How many X-box games does they have all together now ?

(D) Lupin had 11 X-box games. He has sold 5 of them. How many X-box games are left with him now ?

41. What is the variable in the expression 121 p - 35 ?

(A) X (B) p

(C) 121 (D) -35

42. Which of the below options represents a translation of four units left and seven units down ?

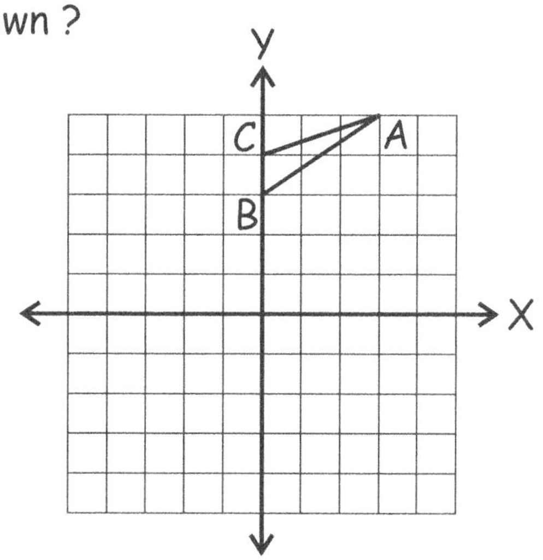

(A)

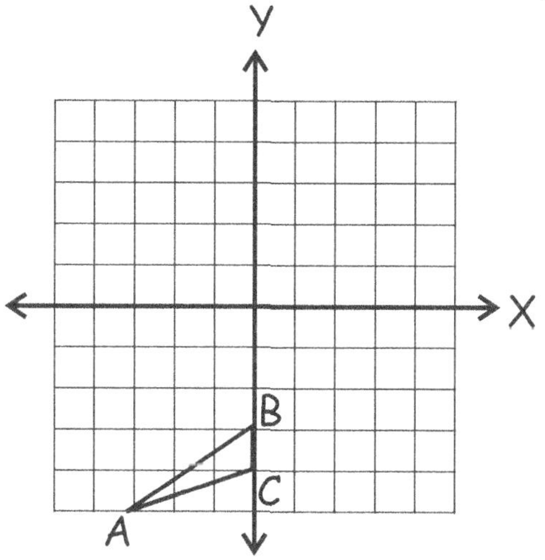

(B)

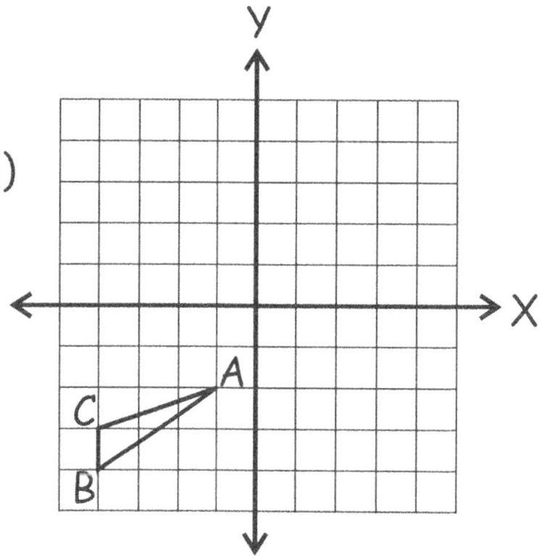

(C)

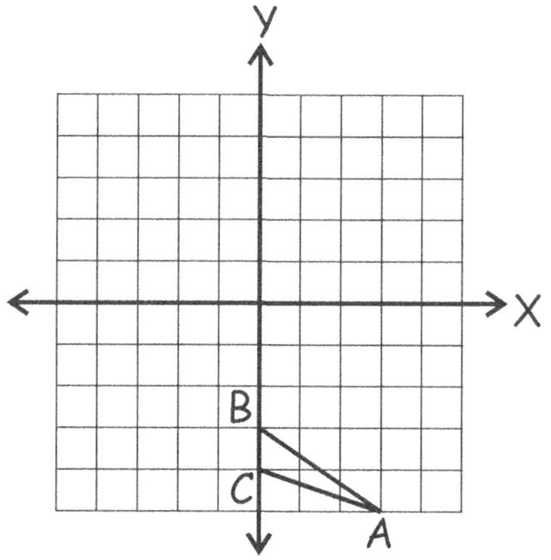

(D)

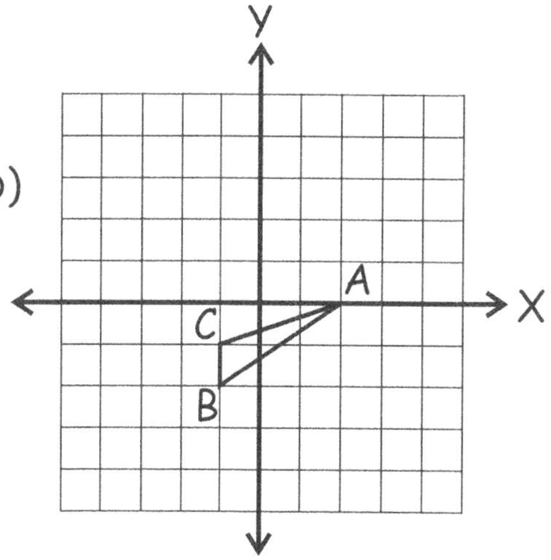

43. Gabby scores 5 less goals than the team score in his soccer game today. If g is the number of goals made by the team. How many goals did Gabby score today ?

(A) $g + 5$ (B) $g - 7$

(C) $g \times 5$ (D) $g - 5$

44. A stationary store sells "e" envelops in one box. Stacy bought 6 boxes today. How Many envelops did she bought in total ?

(A) $e + 6$ (B) $e-6$

(C) $6e$ (D) $\frac{6}{e}$

45. Danny distributes 2262 blankets equally among 6 orphanages, representing the equation $2262 \div 6 = b$. What does b represent in the equation ?

(A) Number of blankets each orphanage received

(B) Number of blankets left

(C) Number of Orphanages

(D) Number of blankets each person received

46. If "s" muffins are distributed among 24 kids in the class. Which of the below expression represents the number of muffins each student receives?

(A) s + 24 (B) s*24.

(C) $\frac{s}{24}$ (D) s - 24

47. Number sentence 5d = 20 can be written in words as

(A) Ethan baked 5 muffins in 20 batches . How many muffins did he bake all together ?

(B) Ethan baked 20 muffins in 5 batches . How many muffins did he bake in each batch ?

(C) Ethan baked 5 more muffins in each of 20 batches . How many muffins did he bake in total ?

(D) Ethan baked 5 less muffins in each of 10 batches . How many muffins does he need to bake more ?

48. William solves three times as many puzzles as his friend Lola. Lola solves 34 puzzles. Which number sentence could be used to find p, the number of puzzles that Will has ?

(A) p = 3 + 34
(B) p = 3 - 34
(C) p = 3 ÷ 34
(D) p = 3 × 34

49. This table shows running time on track by rak each day

Day	Time (hr)
1	0.30
2	0.45
3	1.00
4	1.15
5	1.30

If the pattern continues, What is the time spent by Rak at the end of 7th day ?

(A) 2 hours , 0 minutes

(B) 1 hour , 45 minutes

(C) 2 hours , 30 minutes

(D) 2 hours , 15 minutes

50. Which of the below pair of pictures are symmetric to each other ?

(A)

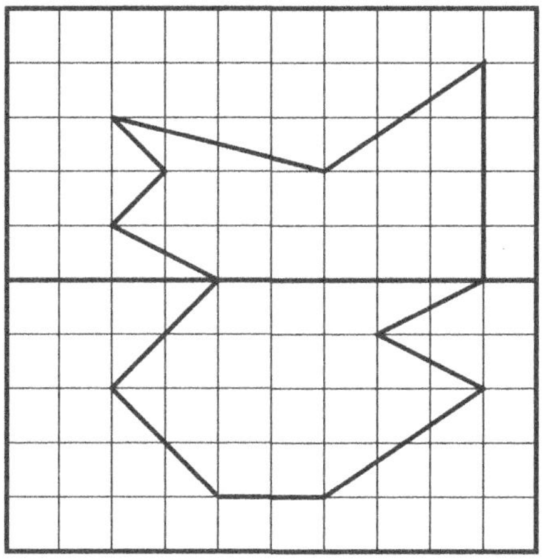

(B)

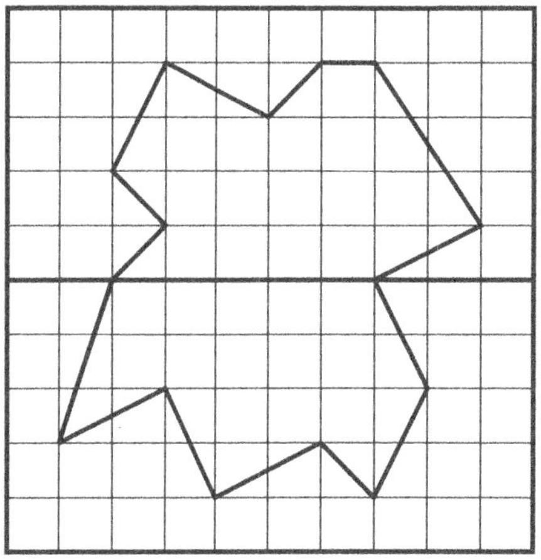

(C)

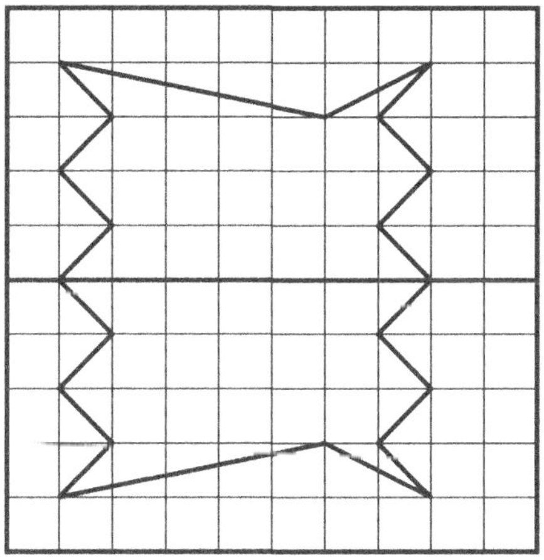

(D)

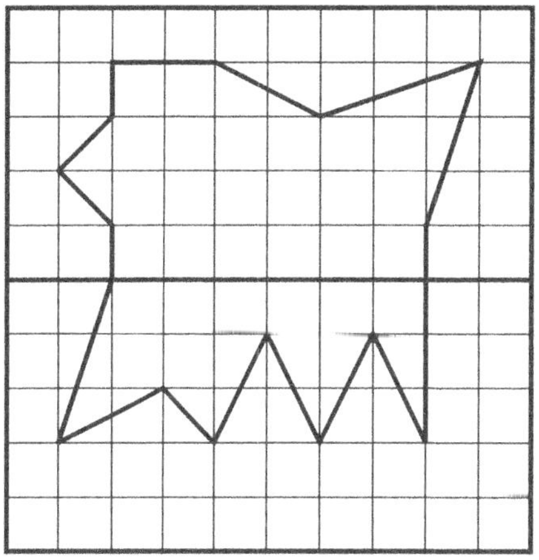

GRADE 5
SOL
Practice Test - 4
Answer Keys

Answer Key Test - 4

1. A
2. C
3. D
4. C
5. A
6. B
7. A
8. C
9. B
10. D
11. A
12. C
13. D
14. A
15. A

16. C

17. D

18. C

19. C

20. A

21. B

22. B

23. A

24. C

25. A

26. A

27. C

28. B

29. A

30. D

31. A

32. D

33. A

34. B

35. C

36. B

37. A

38. A

39. D

40. C

41. A

42. D

43. B

44. A

45. A

46. C

47. D

48. A

49. A

50. B

GRADE 5
SOL
Practice Test - 5
Answer Keys

Answer Key Test - 5

1. D
2. A
3. C
4. B
5. A
6. D
7. C
8. A
9. D
10. C
11. A
12. C
13. D
14. C
15. C

16. A

17. D

18. B

19. C

20. D

21. C

22. A

23. D

24. B

25. D

26. A

27. C

28. D

29. B

30. B

31. D

32. D

33. B

34. C

35. A

36. D

37. A

38. C

39. A

40. B

41. D

42. A

43. D

44. A

45. C

46. A

47. A

48. C

49. B

50. C

GRADE 5
SOL
Practice Test - 6
Answer Keys

Answer Key Test - 6

1. C
2. D
3. A
4. C
5. B
6. D
7. A
8. A
9. D
10. B
11. A
12. C
13. D
14. C
15. C

16. A

17. D

18. B

19. A

20. C

21. D

22. B

23. C

24. A

25. D

26. B

27. D

28. D

29. A

30. D

31. C

32. B

33. C

34. A

35. D

36. C

37. A

38. A

39. B

40. D

41. B

42. B

43. D

44. C

45. A

46. C

47. B

48. D

49. A

50. C

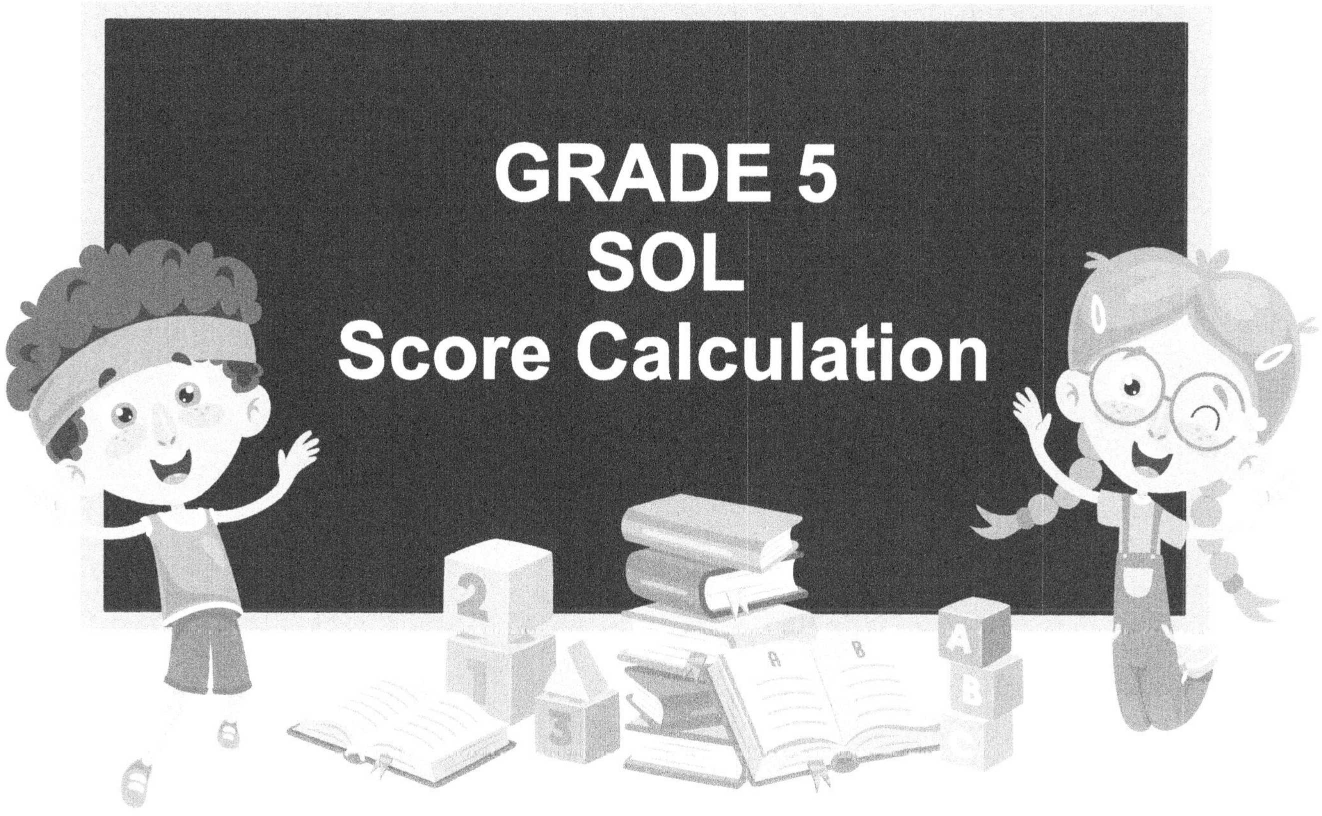
GRADE 5
SOL
Score Calculation

Score calculation

If you get this many times correct :	Then your converted scale scor is :
0	000
1	217
2	249
3	268
4	282
5	293
6	303
7	311
8	319
9	325
10	331
11	337
12	343
13	348
14	353
15	357
16	362
17	366
18	371
19	375

20	379
21	383
22	387
23	391
24	395
25	399
26	403
27	407
28	411
29	415
30	419
31	423
32	427
33	432
34	436
35	441
36	445
37	450
38	456
39	461

SCORE

40	467
41	473
42	480
43	487
44	496
45	505
46	516
47	531
48	550
49	582
50	600

Made in the USA
Coppell, TX
26 April 2022

77090968R00063